T0191682

The Robotics Divide

Antonio López Peláez
Editor

The Robotics Divide

A New Frontier in the 21st Century?

 Springer

Editor
Antonio López Peláez
Universidad Nacional de
Eduación a Distancia (UNED)
Madrid
Spain

ISBN 978-1-4471-5814-1   ISBN 978-1-4471-5358-0   (eBook)
DOI 10.1007/978-1-4471-5358-0
Springer London Heidelberg New York Dordrecht

Printed on acid-free paper

Springer is part of Springer Science+Business Media (www.springer.com)

Contents

1 Introduction... 1
Antonio López Peláez

2 From the Digital Divide to the Robotics Divide?
Reflections on Technology, Power, and Social Change......... 5
Antonio López Peláez

3 The Impact of New Technologies in Organisations........... 25
Áurea Anguera de Sojo

4 Economic Impact of Smart Specialization and Research
in Advanced Adaptive Systems in a Monetary Union.......... 41
Dimitrios Kyriakou

5 Perspectives on Technological Developments
Applied to Robotics................................... 59
Clara Pérez Molina, Rosario Gil Ortego and Francisco Mur Pérez

6 Robotics in Alternative Energy.......................... 87
Raquel Dormido Canto and Natividad Duro Carralero

7 The Future of Smart Domestic Environments: The Triad
of Robotics, Medicine and Biotechnology.................. 117
José Antonio Díaz and M. Rosario Hilde Sánchez Morales

8 Dependency, Social Work, and Advanced Automation........ 137
Yolanda M. de la Fuente Robles and Eva Sotomayor Morales

9 Lessons from the Digital Divide......................... 157
Eduard Aibar

10 Inequalities in the Information and Knowledge Society:
From the Digital Divide to Digital Inequality.............. 173
Cristóbal Torres-Albero, José Manuel Robles and Stefano De Marco

**11 From "Singularity" to Inequality: Perspectives
    on the Emerging Robotics Divide** 195
    Antonio López Peláez and Sagrario Segado Sánchez-Cabezudo

Book Description 219

CV Antonio López Peláez 221

Chapter 1
Introduction

Antonio López Peláez

Throughout history, power has always been associated with technology, to the point that some key historical transitions—from the Bronze Age to the Iron Age—have taken the name of the technology that triumphed at the time. Yet technology is not only the tools or machines we use, it is also about how we organize ourselves, such as the Macedonian phalanx. Since ancient times, humans have fantasized about intelligent, autonomous machines that would permit us to dominate nature (as well as people). This fascination with technology has always gone hand in hand with concerns about the technology possessed by those whom we consider our enemies or competitors, as well as each society's race to gain control of strategic resources and knowledge. Indeed, technology has always been linked to power, and as such has had lasting consequences on social stratification. It is precisely these questions which have provided the departure point for the chapters of this book, and the ten-year research project in which they are framed[1]: the social consequences of technological innovation, and to what extent such innovations affect the distribution of power in our societies.

We find ourselves faced with an intractable paradox: technology plays a key role in our societies. In the initial stages of its development, it is very difficult to establish criteria to guide technology in one direction or another from the perspective of a democratic society as we know little about the costs, opportunities, or positive and negative effects of technology. Indeed, various actors in our societies

---

[1] Since 1995, we have conducted a prospective research project using a combination of methodologies in the field of robotics and other key technologies. Some of the results have been referenced in the OECD Future Studies database, and published in leading journals in the field of technological forecasting such as *Technological Forecasting and Social Change* (López et al. 2012; López and Kyriakou 2008). We are grateful to José Felix Tezanos, Professor of Sociology at the UNED and the driving force behind prospective studies in Spain, for his ongoing support of our research.

A. López Peláez (✉)
Department of Social Work, Faculty of Law, National Distance
Education University (UNED), C/Obispo Trejo 2 28040 Madrid, Spain
e-mail: alopez@der.uned.es

A. López Peláez (ed.), *The Robotics Divide*,
DOI: 10.1007/978-1-4471-5358-0_1, © Springer-Verlag London 2014

compete for and against each specific technology, providing data or exerting pressure to pursue their economic goals. When a technology is implemented in a particular way, there are always winners and losers and people, groups, communities, or countries often remain inside or outside the frontier (the best example is nuclear weapons).

Because the inertial development model in which technology is implemented responds to the current economic and social model of a given society, technological development often reinforces existing inequalities (it is a product that both reflects and reproduces the society in which it is developed). As Collingridge (1980) pointed out, this initially undefined situation contrasts with the final outcome: when a technology has been implemented and reached maturity, it is very difficult to alter its trajectory. And this gives rise to a seemingly unresolvable dilemma: at first, we lack the information needed to act, but once the technology is implemented, it is very difficult to change its course of development.

Awareness of the strategic nature of technology, and hence, science and technology policy, has led to the development of prospective methodologies whose object of study is the possible future in order to analyze the most likely scenarios (where possible) and provide strategic information for decision making. In contemporary societies, we are engaged in an ongoing race to anticipate possible technologies and develop them faster and better than our competitors. This is the origin of prospective research institutes such as the Institute for Prospective Technological Studies (IPTS) of the European Union, and companies such as the Rand Corporation that analyze foreseeable trends and possible scenarios in a wide variety of spheres. In our opinion, analyzing the probable future allows us to open the black box of technological development and describe the potential consequences of such development and the social model in which technology is both implemented and reinforced. In democratic societies, prospective studies provide a space to reflect upon our model of development and the choices we make in our current historical context.

Robotics, in particular, is now a mature technology that has expanded widely in industry in the last two decades (IFR 2012a), and grown exponentially in the service sector (IFR 2012b). Three main areas of debate have arisen around robotics: the now-classic debate dating back to Aristotle on the emergence of an automaton that could be a new human being, a notion which has been redefined by Moravec (1999) or Kurzweil (2005), the debate on the impact of robotics on employment in the industrial and service sector and the complex relationship between technology and employment (Castells 1996) and finally, the changes in human–machine interactions brought about by advanced robotics (Levy 2007).

In this book, we present the reader with a different approach: we inquire into the effects of technology on power from a very specific perspective that arose from a shared experience in recent years relating to the obstacles to accessing new technologies. Could we, in an almost immediate future, find ourselves face to face with a new divide, a new barrier between the included and the excluded arising from the sociotechnological model in which advanced robotics is integrated?

Will we see a new *robotics divide*, which will redefine and expand the impact of the already widely studied *digital divide*?

This question can only be answered by exploring the following issues:

- First, the role technology plays in our economy and in our organizations, and to what extent we can define what we call the robotics divide. To explore this issue, it is necessary to define the approach to be taken and examine in depth the role that new technologies have, until now, played in today's society. To do so, Chap. 2 deals with the basic dimensions of the robotics divide, which is defined as follows: "the distance or separation between those individuals, companies, and states that possess the economic, as well as scientific and technological, capacity and resources to develop robotics technology has redefined their spheres of production and leisure in order to incorporate robots, can make the necessary investments in those spheres, has developed, and has at their disposal advanced robots in the military and aerospace field and the Internet, and those individuals, companies, and states that do not have these resources." The following two chapters examine the role played by technology in organizational dynamics. In Chap. 3, professor Anguera analyzes the changes that have taken place in organizations when implementing new technologies, while in Chap. 4 Dimitris Kyriakou discusses how technologies, particularly robots, can be used to improve the competitiveness of our economies. Given the crucial role played by new technologies in our society, it is essential to address the socioeconomic consequences arising from the various models of technological development.
- Secondly, the different lines of development of advanced robotics and new spheres of application. In this regard, professors Gil, Mur, and Pérez examine the technological development of robotics in Chap. 5. In Chap. 6, professors Dormido and Duro discuss a key area: robots in the field of alternative energy. Professors Díaz and Sánchez then explore the foreseeable advances in medical and domestic robotics in Chap. 7, while in Chap. 8, professors De la Fuente and Sotomayor address key aspects in the development of service robotics in coming years, specifically with regard to people with limited autonomy and mobility.
- After analyzing the role of new technologies in our societies and foreseeable developments, we then turn our attention to the features of the sociotechnical development model in which robotics is integrated into cosmopolitan societies. To do so, Professor Aibar analyzes the characteristics of the digital divide in Chap. 9, while professors Torres, Robles, and de Marco explore inequalities in information societies in Chap. 10. Finally, Chap. 11 is dedicated to future technological trends in the field of robotics in the next fifteen years, particularly applications that can have a strong impact on increasing inequality such as military and space robotics and the development of automated and robotic systems on the network.

In short, we hope to provide the reader of this book with some interesting answers, and, above all, a new approach to the analysis of the human–robot relationship. Undoubtedly, the twenty-first century will be the century of robots (as it already is in industry). For this reason, alongside other reflections on the emergence of a new human alter ego, it is important to reflect on power and inequality. In this regard, we hope that the dreams of a happy society of machines, and the nightmares of a confrontation between robots and humans, can benefit from research on the emerging robotics divide, permitting us to design more inclusive societies in which automatic and robotic systems will contribute to improving people's lives.

References

Castells M (1996) The information age. Economy, society and culture, vol. 1. The rise of the network society. Blackwell, Malden

Collingridge D (1980) The social control of technology. Pinter, London

International Federation of Robotics (IFR) (2012a) World robotics. Industrial robots 2012. IFR, Geneva

International Federation of Robotics (IFR) (2012b) World robotics. Service robots 2012. IFR, Geneva

Kurzweil R (2005) The singularity is near. When humans transcend biology. Viking Penguin Group, New York

Levy D (2007) Love and sex with robots: the evolution of human-robot relationships. Harper, London

López Peláez A, Kyriakou D (2008) Robots, genes and bytes: technology development and social changes towards the year 2020. Technol Forecast Soc Chang 75:1176–1201

López Peláez A, Segado Sánchez-Cabezudo S, Kyriakou D (2012) Railway transport liberalization in the European Union: freight, labour and health towards the year 2020 in Spain. Technol Forecast Soc Chang 79:1388–1398

Moravec H (1999) Robot mere machine to transcendent mind. Oxford University Press, New York

Chapter 2
From the Digital Divide to the Robotics Divide? Reflections on Technology, Power, and Social Change

Antonio López Peláez

2.1 Introduction

Technology is an inseparable part of our lives. While we are ultimately responsible for its creation, there can be no doubt that technology, in turn, is shaping our very existence, and, as with any social institution, it has taken on a life of its own as a consequence of its complexity and evolution. The technology we use on a daily basis is not just hardware; it is also software, the way we organize ourselves, and the tools and machines we develop. Humans, as living beings, are technological tools of the highest level of complexity. And, just like machines, we can be dissected and elements of our organism can be transplanted or redesigned. Furthermore, our machines and organizational systems are not overtly transparent to us; owing to their complexity, they require comprehensive preliminary training (via educational institutions) and financing for development (hence science–technology policies are now a strategic issue in knowledge societies) and have become a core element in explaining power and social inclusion. No other form of social exclusion is more dramatic than death as a result of superior military technology (science fiction films, like the Terminator saga, abound with robots or alien beings that develop weapons of mass destruction). In this chapter, we discuss some key dimensions of our technological society, in order, in a second phase, to investigate the characteristics of what we call the new robotics divide.

A. López Peláez (✉)
Department of Social Work, Faculty of Law, National Distance Education University
(UNED), C/Obispo Trejo 2 28040 Madrid, Spain
e-mail: alopez@der.uned.es

A. López Peláez (ed.), *The Robotics Divide*,
DOI: 10.1007/978-1-4471-5358-0_2, © Springer-Verlag London 2014

2.2 Science, Technology, Society, and Power

In all societies throughout history, science and technology have played a crucial role. We only need to look at the impact of agriculture, writing, or, from a military perspective, the discovery of bronze and iron. Technological innovation has always been linked to the race to obtain technology (at the outset), among the relevant actors in any given historical context. Today, technological innovations shape our power structures and determine the economic, political, and social inclusion or exclusion of individuals, companies, and states. Many of the hidden battles being waged today between countries and between companies have to do with innovation, industrial espionage, and access to key technologies. The difference is that, nowadays, science and technology have developed into a crucial institutional complex that attracts heavy investment and have also become a strategic issue as regards political action. Below, we highlight some features of our current historical context:

- First, we are immersed in a scientific and technological revolution which increasingly affects more dimensions of social life while also acting as a barrier around which new social and economic inclusion and exclusion processes are articulated (López Peláez 2003). In this context, science policies have become a crucial issue for modern societies. Given that "the technological revolution provides the necessary infrastructure for the process of formation of the global, informational economy, and it is fostered by the functional demands generated by this economy" (Castells and Hall 1994: 23), science and technology policies (i.e., those that aim to influence scientific and technological innovation at a specific level of development and orientation via plans drawn up by the political authorities that bring into play their financial, administrative, and educational resources) have become a strategic factor in the twenty-first century.
- Second, from a historical perspective, "(…) the ability or inability of societies to master technology, and particularly technologies that are strategically decisive in each historical period, largely shapes their destiny, to the point where we could say that while technology *per se* does not determine historical evolution and social change, technology (or the lack of it) embodies the capacity of societies to transform themselves, as well as the uses to which societies, always in a conflictive process, decide to put their technological potential" (Castells 1996: 33).
- Third, our culture, our way of life, and our personal and collective aspirations are built on science and technology, as well as the concepts, images, and metaphors that we associate with them. From measuring the speed of cars to learning how to cook, our material culture cannot be separated from our technology, our scientific knowledge. Moreover, since the Renaissance at least, the way we deal with problems is highly influenced by scientific methodology: The scientific gaze gives us the freedom and allows us to transcend tradition and custom in order to respond creatively too and transform the environment. Our

education and science and technology systems contribute to the development of this approach and form a key institutional complex in modern societies.

- Fourth, and as a result of the above, science and technology play a key role in structuring our societies. Indeed, from the moment the first tools were developed by Homo sapiens, technology has always played a key role in human existence. However, it is also true that the importance of technology today, and our own self-knowledge acquired via such innovations as the cell phone, computers, television, or the Internet, makes one wonder about the consequences of the way we design, develop, and implement new technologies and how they contribute to strengthening hierarchical structures or asymmetric distributions of power. And it is precisely this kind of question that inspires fantasy literature to explore the conflict between humans and machines, people and androids, or cyborgs, sentient beings with both organic and cybernetic parts, and robots and humans. If technologies, both old and new, play a key role in the economy, in the production of goods and services, in behavior patterns, in forms of social and political organization, in leisure, and in forms of participation, to speak of a newly emerging technologically advanced society, we should ask ourselves who is included and who is excluded from technological dynamics. That question has been raised in the last three decades in relation to the Internet (the "digital divide"), and now, we need to ask that same question in relation to the technology that seeks to replicate intelligent behavior in autonomous machines: robotics.

The science–technology complex, or "technoscience" as some theorists call it, is one of the key features of emerging societies and responds to and shapes the historical present in a complex process of mutual interaction, recreation, and reciprocal conditioning. Science, technology, and society are closely linked, and the analysis of science and technology as a product of society, in which it is created and which it shapes reciprocally, has evolved from the analysis of its institutional dimension and the impact of science and technology on society, to the analysis of the social construction of scientific knowledge (which also covers the cognitive dimension of science). The two features of the current, which give rise to so-called technoscience, are as follows: first, the exponential acceleration of knowledge production and second, the reduction in the time needed to develop technologies based on scientific advances that can be applied in a practical context. In short, the consequences of the scientific–technological revolution are global, affect more and more dimensions of social life, and are rooted in permanent innovation. Within this context, robotics has an important role, which we aim to analyze in different chapters of this book.

2.2.1 Citizenship and Science and Technology Development

The key role played by technology in our societies has sparked a heated debate about its characteristics, its unwanted side effects, and the governance of technological development. In democratic societies, in which people want to be in control of their own lives and decide how their societies should be run, technology cannot be excluded. Yet technology should not materialize as if by accident, or be the sole responsibility of the experts, leaving us with no other option than that of uncritical acceptance, silence, or, in the best case scenario, passive resistance. Any catastrophe that is precisely a result of technological advances (such as a nuclear disaster, or an environmental disaster resulting from the sinking of a supertanker—it should not be forgotten that navigation and ships are also technology, technical artifacts) highlights the possible paths and consequences of technology. For this reason, forecasting is not only concerned with setting a timeline for potential implementation in forthcoming years, but also with the competitive position of individual countries, as highlighted in the series of Delphi studies conducted in Japan. As regards the military, it is clear that no nation state can afford to ignore innovation, and, in effect, there is hard-fought competition to achieve a position of dominance.

In democratic societies, debates on science and technology development cover all areas: from questioning the source of science funding to discussing the design, features, and impact of science and technology. Simultaneously, this has been accompanied by a significant increase in research into scientific and technological practices, in what has been called the new "sociology of scientific knowledge," in which not only the impact but also the invention, development, and internal characteristics of science and technology are considered social facts/processes. This research has led to what is now referred to as the "Science Wars" debate (Ross 1996). In a parallel manner, a whole line of research has been developed on the public perception of science and technology in knowledge societies, in which citizens demand to be treated in accordance with their status in all spheres of life and not reduced to mere passive objects when it comes to science and technology policymaking (López and Díaz 2009).

The future evolution of the scientific–technological revolution returns us to the debate on the social models in which it is created and implemented and which, in turn, transforms and shapes. From this perspective, prospective studies on social trends aim to provide qualified information for decision making (López Peláez 2009). In this respect, we cannot analyze the processes of social inclusion and exclusion without considering the technologies that will play a critical role in the coming years. And, from our point of view, robotics will radically transform both our environment and our way of life (Kurzweil 1999). We need information in order to make decisions, and it is precisely the effects of technology on social exclusion, as well as the barriers it creates by redefining what is included and what is excluded from the core of our economy and our way of life that we need to analyze when contemplating how we are going to integrate our own robotic designs into our democratic societies.

The "runaway world" that Anthony Giddens speaks about is undoubtedly a rapidly changing world in terms of science and technology. Although both are products of their social context, from our point of view, there is an important feature that ought to be highlighted: Once stabilized and implemented, the technological advance in question determines future developments while simultaneously constraining and enabling our field of action depending on what it is exactly achieved through such technology. In this regard, Thomas Hughes' (1983) analysis of what he calls "sociotechnical systems" (consisting of physical artifacts, organizations, natural resources, legal mechanisms, and intangible components of organizations, etc.) highlights how such systems evolve from the interaction between economic resources, practical skills, and organizational forms. Once the system is firmly established, it tends to remain stable until the political and economic interests linked to the system are exhausted.

From a different perspective, the analysis of technological development and its social impacts reveals the sociopolitical dimension of technology, which is built on and responds to a specific social model. Technological artifacts can be designed and constructed so as to produce a series of logical and temporal consequences prior to a concrete application. In this regard, technologies can be described as "life forms" that establish ways to build "order" in our world. They hold various possibilities for organizing human life in various ways. Winner (1987) highlights three key features of the process of technology creation and acceptance:

- Consciously or unconsciously, societies "choose" technological structures that have a long-term influence on the way they work, communicate, and live.
- In the decision-making process, people occupy unequal levels of power and awareness.
- The range of choices is greater when technology is introduced for the first time. Once implanted, it establishes a pattern of social order that is very difficult to change. "Because choices tend to become strongly fixed in material equipment, economic investment, and social habit, the original flexibility vanishes for all practical purposes once the initial commitments are made" (Winner 1987: 45).

From Winner's point of view, "What appear to be merely instrumental choices are better seen as choices about the form of social and political life a society builds, choices about the kind of people we want to become" (Winner 1987: 69). From our perspective, technology should be seen as a social fact/process, in which both the social interests that drive their development, and the logic arising from the interaction between technological revolution, capitalist restructuring, and social changes can be traced. Technological revolution penetrates all spheres of human reality. The relationship with the social contexts in which it is created, applied, and which it transforms in often unpredictable ways for its creators, requires a scientific approach that contemplates two aspects: social processes that give rise to the technological society in which we find ourselves and the logic arising from technological developments (and their specific social application), which acts upon and transforms the social reality.

The impact of a new technology is never limited to the purpose for which it was designed. "Technology also implies a transcendence of the materials used to comprise it. When the elements of an invention are assembled in just the right way, they produce an enchanting effect that goes beyond the mere parts. (...) The assembled object becomes much more than the sum of its parts" (Kurzweil 1999:

33). And this applies not only to the materiality of technological artifacts, but also to their social application. Hence, the unanticipated side effects and the unexpected path are taken by technological development for those who design, fund, and implement technology.

Models of social development and interpretation of the world that articulate and provide symbolic support to advanced technological societies—depending on the characteristics of both the technological systems and social interests—have become the subject of sociological analysis of technology. This leads us to the processes by which the "modern image" of society and the world are built. This modern image has become the symbolic universe that has guided technological developments and justified their implementation via the notion of "progress." At the same time, the advanced technological society that arises from this development calls into question basic social consensus and how the previous society was organized: "(…) industrial society *exits the stage of world history on the tip-toes of normality,* via *the back stairs of side effects.* (…) the counter-modernistic scenario currently upsetting the world (…) does not stand in contradiction of modernity, but is rather an expression of reflexive modernization beyond the outlines of industrial society" (Beck 1998: 17, emphasis in the original). In this process, the paradox of the undesirable and unforeseen consequences of scientific and technological development is revealed, forcing us, as Beck puts it, to deal with "the growing capacity of technical options" alongside the growing "incalculability of their consequences" (Beck 1992: 22).

In short, in what follows, we highlight some features of our social model of science and technology development, in which advanced robotics are designed and integrated.

- First, we are immersed in a permanent scientific and technological revolution, which began with the information and communication technology revolution in the 1970s. It is a revolution characterized by the accelerated pace of science and technology developments and converging technologies.
- Second, the unknown effects of the rapid expansion of science and technology in various fields of social and natural life. From genetic engineering to artificial intelligence, new risks and opportunities are transforming our environment, and we face the problem of the undesired consequences of our actions (with a level of "new" radicalism, given that what is at stake is the very survival of the human species). This perspective paved the way for technology assessment as a strategy to democratically redefine the sociotechnical model we want for tomorrow's society. In this process, the social nature of scientific and technical output is revealed in three dimensions:

  – It responds to a way of life that facilitates, as the Spanish philosopher Ortega y Gasset so brilliantly put it, the recreation of an artificial world, in which a future imagined by humankind is made possible.
  – Science and technology development responds to the interaction between various social groups and interests. As such, neither technological determinism (in its "hard" version, which implies the existence of a technological

logic independent of the society in which the technology is created and applied) nor the uncritical identification between sociotechnical progress and development can be defended: The hypothetical "neutrality" of science and technology does not respond to the reality of science and technology as complex human activities.

– Sociotechnical developments constrain the present and the future in such a way that interaction between science, technology, and society must overcome both technological and social determinism.

• Third, societies based on technoscience are characterized by anticipating the future, which becomes the benchmark that determines action by developing strategies to reduce uncertainty and ensure not only the economic but also the social viability of the future. The need to decide socially about the future of our societies by evaluating technologies has led to the development of prospective methodologies that permit forecasting the characteristics of the immediate future while bringing to the fore the consequences of today's technoeconomic development model (López Peláez et al. 2012). It should not be forgotten that access or not to certain technologies can have a decisive impact on the future: a fact constantly highlighted in the military sphere.

2.2.2 Opening the Black Box of Technological Development: Forecasting and Assessing Technology

The relationship between science, technology, and society in the current historical context has three characteristics that have led to the development of prospective methodologies linked to predicting technological events and social forecasting, which is concerned with social impacts and seeks to provide qualified information for decision making. These three characteristics are as follows: firstly, the speed of development and the new risks associated with the implementation and management of technology, their design, and ultimate goals (as in the case of genetic engineering, robotics, or nanotechnology); second, the impact of new technologies on labor (the primary means of social integration in salaried societies) and political institutions, primarily on the role of the nation state; and third, the relationship between technological revolution, informational capitalism, and the environment, specifically the need to develop environmentally friendly technologies, which was already highlighted in the debate on sustainable development models as early as the 1960s.

The link between unconditional optimism and science and technology development reached its peak in the 1940s and 1950s when scientific and technological progress was seen as the key to material progress and military supremacy and became a strategic activity that merited unconditional funding from public institutions. Having said that any unconditional funding ought to respect the self-regulatory code of conduct adopted by the scientific community, without interfering in its creative work. The process that led science and technology to become the subject of public debate in the 1960s, and the link between technological progress and scientific development to become problematic, was due to the interaction of several factors: first, the impossibility of providing unlimited resources to fund all the R&D proposals of the scientific community; second, the strategic role that science and technology research played in economic development, which made it the object of political debate; third, the potential risks of technological advances such as nuclear disasters; fourth, the limits of industrialism and technologies based on environmental depredation and the depletion of non-renewable natural resources; and finally, the systematic criticism of technocratic instrumental rationality and the debate on the epistemological status of scientific knowledge. All of this triggered reactions in three spheres: the social sphere (demands for public participation in assessing the impacts of science and technology and demands relating to the regulation of the implementation of technological advances), the political sphere (more interventionist policies were adopted, which sought to obtain results from the strategies devised by political powers), and in the field of social sciences. The end result was the transformation of the implicit contract between science, technology, and state, and the design of a new social contract characterized by the abandonment of technological determinism, the pursuit of citizen participation, and the assessment of the impact of technology developments.

According to these approaches, research on the relationship between technological change and social change should address three key issues: first, the debate on the influence still exercised by technological determinism today; second, the role of science and technology in today's society, both in terms of its institutionalization and the plans for its design, impact, and assessment, and in terms of its public perception of scientific and technological advances; and third, the analysis of science and technology as social facts and processes and hence research into its impact and own internal structure (in the context of both discovery and justification). In our opinion, the secularization of science and scientific activity, which turns it into an object of sociological inquiry, is another step in the process to define knowledge and action as activities inherent to all human beings.

Despite the siren song of technological determinism, and its corresponding sociohistorical determinism, human beings have always been concerned with the future. They have always questioned the consequences of their actions and how to intervene in the likely development of events. The current debate on the relationship between technological change and social change has been repeated many times in the past. The future has always been a cause of concern for humankind. Throughout history, we can trace the many ways of analyzing the changing trends recorded in the present and how to predict the characteristics of the future. From the oracles in classical antiquity (Hernández de la Fuente 2008), to modern prospective studies (Fowles 1978; EFILWC 2003; Georghiou et al. 2008), action in the present cannot be understood without taking into account the objective to be reached within a set period of time.

The basic difference between the various forms of prediction employed throughout history lies precisely in the notion of future with which they operate. In ancient times, and to this day, traditional determinism (technological, historical, social) postulates a single rationale for the development of events in which human subjects are passive recipients of change and are responsible for that change only to the extent to which the process is said to be in their favor (hence the various rites to ask whether the gods are on our side). This model brings us to the Heideggerian analysis of what lies beneath the essence of technology; what Heidegger (1975) describes as *das Gestell*, in which there is no other option other than to wait for a "new advent." In this extreme approach, the future is also determined by inherent laws beyond our control, which take technological determinism to its maximum expression.

This conceptualization of the future from the perspective of determinism entered into crisis after World War II and the development of the so-called Big Science. Big Science saw the future as a playground for human action: a place to be conquered where passivity is no longer an option, and as a result, methods and techniques were needed to analyze and build the future according to political, economic, and military interests. Technological objectives such as the atomic bomb required years of collaborative planning and work, the provision of substantial resources, and a clear assessment of the benefits to be had (which are gained not only from obtaining key technology in military or civilian fields of expertise, but also by assessing the costs and risks of other countries or companies obtaining technology sooner or improving it).

In this new context, technological forecasting became a strategic tool and was soon regarded as a key element in R&D programs in technologically advanced countries. In uncertain environments characterized by strong competition between companies and countries and the need to invest financial and human resources (which are always limited) in technologies that are regarded as key in the immediate future, technological forecasting provides "difficult-to-acquire strategic information for decision making, and it functions as an socioeconomic mobilization tool to raise awareness and to create consensus around promising ways to exploit the opportunities and diminish the risks associated with S&T developments" (Morato et al. 2002: 5).

In advanced technological societies, R&D policies demand qualified information about the future in order to develop short-, medium-, and long-term plans for three main reasons: first, the need to anticipate the characteristics of the technological model within a time frame of 10–15 years, e.g., establishing an energy production, distribution, and consumption model in a country, requires planning, design, construction, and the implementation of a group of production facilities and distribution networks based on the forecasts made; second, errors in forecasting the future immediately give rise to serious problems: "(...) failure to anticipate subtle cumulative long-term changes precipitate sudden short-term consequences (...)" (Abt 2003: 88); and third, the need to establish a continuous reevaluation process of the forecasts and results to adapt to changes in the environment: "The goal becomes making the most adaptive decisions in a timely fashion rather than getting the future right" (Schwartz 2003: 37).

Planning entails forecasting, evaluating, selecting priorities, and making decisions based on the available resources. Debates on the limits of our methodologies for forecasting the future should not make us forget that we are immersed in a scientific and technological system in which we are funding innovations that will be on the market within 10 years and that the ability to cope with present and future risks will depend on our strategic planning and on how we act now to address possible future risks. It should therefore come as no surprise that future studies were developed in the 1950s during the Cold War in the military sphere and in a context of technological innovation (nuclear energy and the space race) that could have had an impact on the existing balance of power. Currently, most of the EU member states, and many countries in other regions of the world, conduct systematic forecasting studies in order to provide qualified information for decision making in the field of science and technology policy.

In democratic societies, risk analysis also has other direct consequences: People want more information and greater participation in the development and implementation of technologies that affect their daily lives. Technology forecasting cannot ignore the social consequences of future developments and the key role played by citizens as consumers (in the case of products) or policymakers (in the case of technologies that affect the definition of our identity as human beings). As a result, the methodologies used in future studies increasingly include analyses on social impacts, thus increasing the amount of information available for decision making.

As highlighted by the so-called Collingridge dilemma (Collingridge 1980), this becomes more relevant due to the fact that it is more difficult to make decisions to guide or shape technology in the initial stages of development given that we know little about its costs, opportunities, risks, and positive and negative effects. On the other hand, once technology has been developed and implemented, and we do have enough knowledge about it, it then becomes very difficult to alter its trajectory. One possible way to overcome this dilemma is to provide, via prospective methodology, adequate information before technological trajectories become irreversible. As a result, prospective studies have become a strategic tool for the internal competitiveness of countries and for the democratic system itself, to the extent that the technology choices we make, or fail to make, often shape our lives in a definitive manner.

In the new context of the relationship between science, technology, and society in democratic societies where citizens are seen as subjects and not objects and therefore demand greater participation in building their own future, public demand to participate in technology assessment has increased. In a parallel manner, the demand for information has increased as regards both the current consequences of scientific and technological development and the diverse possibilities for the future and the forecasts relating to decision making. Finally, a wide field of study has been developed on public perception of science and the nature of scientific knowledge, as well as the interaction between science, technology, technical artifacts, and human action. Using an interdisciplinary perspective, our study will focus on a key technology that will exert a major influence on our society in the future: robotics.

2.2.3 From Technological Determinism to a New Social Contract Between Technology and Democracy?

The debate about the influence of technology on society as an agent of change can be analyzed in terms of the debate on what is known as "technological determinism." The enlightenment project is based on the notion of human's capacity to learn by shaping reality, and as such, every human product is historical and is produced, and responds to, a specific logic of action. It is for this reason that the enlightenment aimed to emancipate human consciousness from of all kinds of external determination and superstition and subject it to a "coming of age" based on the premise of rational behavior.

In this process, science as a knowledge system plays a critical role: It enables the advancement of knowledge, transforms nature, and facilitates a new order derived from the combination of knowledge, freedom, and usefulness. In this context, the legitimacy of science and the practical use of technology lead progress to be regarded primarily as scientific and technological progress that allows us to change our way of life and create a better society. Furthermore, technical artifacts, once introduced, seem to take on a life of their own, to the point that subsequent events tend to be explained as an inevitable consequence of technological innovation. "A sense of technology's power as a crucial agent of change has a prominent place in the culture of modernity. (...) For some three centuries, direct firsthand experience of that power has been a well-nigh universal feature of life in developed and developing countries" (Smith and Marx 1996: 11–12).

Moreover, "The idea of technological determinism takes several forms, which can be described as occupying places along a spectrum between 'hard' and 'soft' extremes. At the 'hard' end of the spectrum, agency (the power to effect change) is imputed to technology itself, or to some of its intrinsic attributes (...) At the other end of the spectrum, the 'soft' determinists begin by reminding us that the history of technology is a history of human actions" (Smith and Marx 1996: 14–15).

The key issue, in our opinion, is not only the social origin of technology, but also its effectiveness once developed, as well as its limitations and opportunities. In this regard, the debate on the internal logic of science and technology development, as highlighted by Mumford, Ortega y Gasset, and other early twentieth-century scholars, leads to the model of social organization in which technology is developed and which, in turn, is reinforced and modified by it. As a historical phenomenon, technology is contingent in origin, yet at the same time, as a historical phenomenon, it is also a part of the trajectories that condition the future. In the works of Mumford and Ellul, technology is not understood as mere machinery, but also as a form of organization, and a whole manner of thinking rooted in efficiency and calculation.

In this regard, their pessimism ties in with Heidegger's critique of technology for whom the essence of technology is not technological at all but a way of life embedded in the language he calls *das Gestell*, or framing, in which everything is reduced to pure commodity and instrumental calculation, and where human identity as a concept is dispersed in the process. In contrast to these negative conceptions, yet placing technological logic beyond the notion of technoscience, Ortega y Gasset adheres to a universal model of thought and action to determine society, proposing technical analysis as a strategy to produce the artificial worlds that typify the human condition as opposed to the natural world. By definition, "possible worlds" are infinite, and as such, many "techniques" which characterize specific sociohistorical projects have been developed throughout history.

These perspectives on the relationship between science, technology, and society are a forerunner to what we believe is key in understanding the interaction between science, technology, and society today:

• On the one hand, the breaking of the covenant between science and the state, which has led to new methods of evaluation, the incorporation of prospective methodology to provide critical information in decision making, and the development of science and technology policies and strategic tools to ensure the economic and social integration of emerging societies, as well as a line of research that investigates public perceptions of scientific and technological progress and impacts.

• And on the other hand, the analysis of science and technology as social products in an area of research that has come to be known as "science, technology, and society." From a strictly sociological perspective, it is important to mention the fields of study that focus on science as an institution, which evolved from the Mertonian school of thought, as well as the so-called new sociology of scientific knowledge, which regards social facts not only as the application and impact of technologies, but also as their process of gestation, design, and modeling.

Modernity has been built on the confidence in the link between scientific knowledge and material progress. This link stems from the utilitarian dimension that characterized both the scientific revolution and the technological evolution linked to the industrial revolution. Although the practical application of scientific progress has not always followed a linear process, examples such as the competitions held to develop technologies and devices in order to accurately measure latitude at sea in order to locate the exact position of a ship and its course clearly demonstrate the link between scientific knowledge, technology, and practical application. The legitimation of science and technology as an engine of progress has been brought about by both the degree of knowledge it provides and the degree of knowledge and well-being embedded in objects, tools, and technologies used in everyday life.

Public ambivalence toward the consequences of technological development is due to the emerging risks and the analysis of technoscience activity as a social process, which is not solely governed by the search for truth and benefits to humankind, but responds to a complex set of interactions between interests of various kinds and the crisis of the scientific paradigm as a cognitive paradigm, as the standard of truth. However, this complex process contrasts with technoscientific advances in daily life. New technologies are projected onto our self-understanding of reality, to the point of becoming, in the words of Bolter, "defining technologies," which are used as metaphors, models, or symbols around which thoughts and actions are organized. From this perspective, the relationship between technoscience metaphors, language, thought, and individual and collective action should be highlighted. Metaphor, as a model for understanding reality, conditions and facilitates our individual and social and even scientific perception of the world. Therefore, the analysis of metaphors becomes an essential tool in order to expand and structure our knowledge and action.

The reliance on technology, and the rapid introduction of all kinds of innovations in everyday life, raises suspicion and causes a certain degree of apprehension (we can see both extremes in a paradigmatic way in contemporary films about robots, from The Terminator to I, Robot). This complex view of the relationship between technoscience progress and development stems from the interaction of several factors:

- First, the impossibility of providing unlimited resources to fund all R&D proposals by the scientific community.
- Second, the strategic role of science and technology research in economic development, which makes it an object of political debate.
- Third, the potential risks of technological advances such as nuclear disasters.
- Fourth, the limits of industrialism and technologies based on environmental depredation and depletion of non-renewable natural resources.
- Fifth, the social divide caused by technology that redefines the boundaries of social inclusion and transforms the context at an individual, company, and country levels. In this regard, there is a strong debate on the positive and negative effects of new information and communication technologies and globalization on several important realms of society such as the economy, the welfare state.
- And finally, the systematic criticism of technocratic instrumental rationality and the debate about the epistemological status of scientific knowledge.

As noted above, this has elicited a reaction in three spheres: the social sphere (demands for public participation in assessing the impacts of science and technology and demands relating to the regulation of the implementation of technological advances), the political sphere "the old laissez-faire policy, which regarded the regulation of science and technology innovation as a matter of internal corporate control, is transforming into a new more interventionist policy where public authorities develop and implement a series of technical, administrative and legislative tools for channeling technoscience development and monitoring its effects on nature and society" (Albornoz and López Cerezo 2007: 45), and in the field of social sciences (as reflected in the research conducted by the so-called new sociology of scientific knowledge, the sociology of technology, and the philosophy of science and technology).

The ultimate consequence is the transformation of the implicit contract between science, technology, and state, and the design of what can be called "a new social contract for science and technology." Social change and technological change are closely related. We cannot take a passive stance and accept technological determinism as an intrinsic law of historical evolution. Technology is a social product and can be built and reoriented according to social consensus. In fact, in the process of being developed, technology already reproduces the basic social consensus of advanced capitalist societies, as well as its patterns of inequality. In order to further consolidate democratic citizenship in the twenty-first century, we need more and better information on scientific development and its potential impacts, greater scientific literacy, and more and better prospective studies that allow us to anticipate the opportunities and risks of key technologies, such as robotics. We must move toward a new social consensus between science, technology, political power, and citizens in order to jointly build a democratic society in which new opportunities permit improving our standards of living and lifestyles.

2.3 From the Digital Divide to the Robotics Divide

Genes, bytes, and robots are going to transform our social life. Our bodies will be redefined through genetic enhancements and more technology, until eventually evolving into cyborgs. Simultaneously, robot intelligence and capabilities will introduce an alter ego in our social environment, and in the same way, they already play a central role in performing tasks in industrial sectors. And all of this is going to materialize in cosmopolitan but unequal and stratified societies, which will possibly create new forms of inequality that will superimpose themselves on older forms of inequality. At the same time, there will be new opportunities, and, to some extent, the worlds of work and entertainment will be redefined and possibly even the world of intimate relationships, including sexual relationships (Levy 2007). The convergence of biology, robotics, and artificial intelligence suggests, therefore, a threefold horizon of possibilities: human genetic enhancement in order to improve our capabilities and make way for a new stage in the evolution of the species; the development of enhanced artificial intelligence until finally producing autonomous machines and robots capable of improving and repairing themselves, which will give rise to sentient beings that are more intelligent than humans; and a mixed future, in which machines such as nanorobots coexist with biological improvements to the human brain, producing a mixture of information and communication robot–human technologies, which will also permit us to take a new leap in the history of life on earth and beyond.

2.3.1 What is New in Robotics?

Twentieth-century literature and film reflect some of the social representations of the potentially negative consequences of progress, of which three merit particular attention: first, the modification of human life and the altering of the species, as shown in works that criticize the eagerness of scientists to modify human nature (as represented in the figure of Dr. Frankenstein or in the current debates on cloning and producing "designer humans"); second, the undesirable consequences of the misuse of science and technology, as can be observed in works that criticize the misuse of nuclear energy. Associated with immoral use, such technologies can produce negative consequences that call into question the viability of the human species and even the very essence of life on earth; and third, the fear of out-of-control technoscience, which, by developing its own potential, becomes independent from human beings; first, creating a way of life governed by the demands of the technological systems themselves to then creating intelligent machines that confront and defeat human beings (as seen in The Terminator, 2001: A Space Odyssey, and The Matrix).

Technophobes and technophiles share a common assumption: Technology, once implemented, shapes the future in a sometimes unexpected but always consistent way. Choosing a technological trajectory is not a trivial matter, and the consequences can be positive or negative for a very long period of time. From a sociological perspective, technologies can serve to transform and recreate nature and human life, or interpret reality and express social conflicts and consensus. And, in times of such rapid technological innovation as the present, reflective uncertainty about the impact of technology is paradoxically linked to the uncritical, rapid, and massive incorporation of all types of artifacts and technologies in everyday life. The technophobe or technocritic in theory and the technophile in practice would seem to be the most common typology in modern societies. In this environment, in which public concern about the health risks of technological innovation is linked to the massive and rapid incorporation of every artifact or technology to hit the market (López Peláez and Díaz 2007), it becomes even more relevant, if possible, to ask the classic question of any prospective analysis: What will be the effects or foreseeable impacts of new technologies in the coming years and how can their consequences be addressed?

The interaction between humans and machines, and especially the possibility of developing robots that can reproduce human behavior and intelligence, is a fascinating field of research. Not only do we transfer our capabilities to robots, but we are increasingly inclined to implant technology and machinery in our own bodies, to the point of making robots more like humans than machines and transforming humans into machines that are more like robots, i.e., cyborgs. From the perspective of the so-called NBIC convergence (Roco and Bainbridge 2002), new possibilities have arisen for transforming the human species by way of a post-modified human life, in which artificial intelligence, genetics, and robotics will make way for a new stage in the history of humankind. The debate on the emergence of a new species has sparked heated discussions in leading forecasting journals (Technological Forecasting and Social Change 2006: 95–127). Moreover, the new models of social relationships that will be established within this context, in which the frontiers between humans and machines will be blurred, highlight the extent to which the current technological trajectories and social models in which we design and implement robots will shape the immediate future in an increasingly irreversible way (as has happened in the past with other crucial technologies). Not only do we need to consider the emergence of a new species, but also the development of a new social subject (as robots are also designed to work as a team, learn together, and interact socially with humans and other robots) in an ontological way. Recent research shows the emergence of a new colleague for purposes of leisure and entertainment, or a partner for romantic and sexual relationships. Given that technology is developed and implemented uncritically and according to market demands, it would come as no surprise that while we are debating on the rights and obligations of robots, they will quickly colonize the sphere of our private lives and become much more to us than a videogame or a pet, but perhaps even romantic partners.

The analysis of probable technological trajectories therefore needs to focus on the social models from which such trajectories are built, always bearing in mind that they could change unexpectedly for those responsible for them. Sociological analysis is precisely what allows us to understand the different possible scenarios arising from expert forecasting in the field of robotics. Increasingly intelligent robots, the increasing robotization of human bodies, and the new possibilities to solve major problems that threaten us as a species (climate change, pollution, the conquest, and colonization of outer space, etc.) all allow us to envisage what Kurzweil calls "the new singularity": a new post-human, post-machine subject. However, if we consider not only the convergence of technologies, but also their practical application in line with the criteria of market societies and the prevailing values of cosmopolitan societies, a key question arises: Will the stratification and segmentation of capitalist societies be reproduced? Are we facing a new divide, the robotics divide, which transcends the current digital divide?

And as to technologically enhanced cyborgs and humans with increased capacity and power, will they be freer, or will they, in reality, reproduce the behavior patterns engraved in the technology they incorporate? According to Norbert Wiener and Joe Weizenbaum, pioneers in the field, implanting integrated circuits in the brains of animals to govern their behavior, could also lead to designing systems to control human beings, where cyborgs would just be robots whose design is based on the human body. Today, we now know about the experiments performed with the so-called ratbot; a rat with electrodes implanted in its brain, thanks to which its behavior can be governed via radio control (Balaguer and Dormido 2007: 490–494). In short, the analysis of future technological trends brings us face to face with society's socioeconomic logic, our relationship with the environment, our behavior patterns, and how we use our power and contributes to opening not only the black box of robotics technology, but also the black box of our technological society.

2.3.2 Approaching a Definition of the Robotics Divide

The term digital divide refers, in particular, to how technology, and access to it, redefines the power structure in contemporary human societies. The questions we pose in this collective work are key for the twenty-first century. Will robotics, which is characterized by converging technologies, rapid progress, and a progressive reduction in costs, be incorporated into our societies under the current market model? Will it create a new technological, social, and political divide? Will it transform power relations between individuals, groups, communities, and states in the same way as crucial military technology? Will it change our everyday social interactions by forming part of our daily lives in the same way it has changed industry where industrial robotics is already mature and fully established? These are crucial questions, because, as we have seen in the past with other key technologies in the Western societies, technology is not neutral, and its foreseeable impacts arise from the socioeconomic models in which we are immersed. In addition, asking these questions obliges us to do two things: first, to provide data on foreseeable trends, risks, and opportunities in line with the best practice of technology assessment; and second, to provide data to enhance our scientific and technological literacy and, with it, our ability to intervene in decision making (whether political decisions via participation and representation mechanisms established in democratic societies or decisions as consumers, which have a crucial effect on the demand for goods and services).

In the following chapters, we present some results on the emerging robotics divide. In our opinion, all technology is implemented in an unequal way, owing to both its cost and usage requirements and the fact that new technologies are almost exclusively used first by what is referred to as "early adopters" to progressively expand into ever-wider layers of the population. In a parallel manner, the type of technology designed depends on the needs it aims to satisfy and the logic that lies behind every technology application, thus causing a set of frequently unanticipated, and often unwanted, side effects. Over a period of 15 years, we have conducted an ambitious prospective research project in the field of new technologies and in the area of advanced robotics in particular. From the results obtained, we have defined a new scenario: the *robotics divide*. In defining this new scenario, we have taken into account the following four dimensions:

- First, the economic as well as science and technology resources that are needed to develop robotics technology in all areas.
- Second, the ability of companies, users, and civil society to reorganize in order to increase economic productivity and incorporate industrial and service robotics into a wider number of spheres.
- Third, the market economy model and distribution of existing resources in advanced societies.
- Fourth, areas in which robotics technology entails redefining power in relation to military and space programs, as well as in the Internet (in which so-called

intelligent agents play a crucial role in accessing goods and services, which are increasingly only available online).

Given these dimensions, we can define the robotics divide as follows: *the distance or separation between those individuals, companies and states that possess the economic, as well as scientific and technological capacity and resources to develop robotics technology, have redefined their spheres of production and leisure in order to incorporate robots, can make the necessary investments in those spheres, have developed and have at their disposal advanced robots in the military and aerospace field and the Internet, and those individuals, companies and states that do not have these resources.* This distance or separation implies higher levels of economic, military, and technological power for those individuals, companies, and states that possess robotics technology and especially in critical areas such as aerospace programs or military combat robots that could gain a competitive advantage which would significantly alter the balance of power between one country and another. As a result, military robotics and aerospace robotics, along with developments in robots for the Internet, will become a strategic issue affecting competition between countries, especially between those that are set to play a leading role in the twenty-first century such as the United States, China, India, and Russia.

In leisure and domestic life, as well as health care and the care for disabled or dependent individuals, new service robots will provide a competitive advantage and could have a significant impact by reducing the need for immigrant workers (spheres of activity which are currently very labor intensive). To a certain extent, the robotics divide could introduce a new actor into our social life, the robot, which could become an assistant, a romantic partner (Levy 2007) or even a warrior, and while robots may not alter job creation in absolute terms, they could have a significant impact on some labor-intensive sectors (i.e., the immigrant workforce in the Western countries). New technologies, in particular robotics, could become a new factor to take into account when analyzing immigration flows (López and Krux 2003). In a more personal context, and to some extent one that runs parallel to the so-called digital divide as noted in some of the chapters in this book, service robotics could become a crucial technology in terms of both access and user requirements. The difference between having and not having a robot may become not only a visible sign of socioeconomic status, but also a symbol of power and wealth, thus creating a gap between the technology "haves" and "have nots" in the sense of Bourdieu. Limited access to robots could also be a clear predictor of social exclusion as they could perform many tasks that would greatly ease the burdens of daily life (not only in terms of mobility or cleaning, but also in terms of emotions and relationships). Table 2.1 summarizes some of the features of the emerging robotics divide at three different levels: states, companies, and individuals.

Table 2.1 Consequences of the robotics divide in the twenty-first century

States	Access to advanced robotics technology	Economic growth and enhanced productivity
		Greater military power
		Border control
		Technological innovation
		Conquer space
	No access to advanced robotics technology	Lower economic growth and productivity
		Less military power
		Less border control
		Lower level of technological innovation
		Increasingly distanced from the space race
Companies	Access to advanced robotics technology	Higher productivity
		Higher levels of automation
		New business niches
	No access to advanced robotics technology	Lower productivity
		Lower levels of automation
		Lower competitiveness in new business niches such as aerospace
Individuals	Access to advanced robotics technology	Automation of domestic chores
		New forms of leisure and services
		Better employment opportunities
		More educational resources associated with robots in the classroom and at home
	No access to advanced robotics technology	Greater difficulties for the disabled
		More activities related to domestic chores
		Less educational resources

References

Abt CC (2003) El futuro de la energía desde la perspectiva de las ciencias sociales. In: Cooper RN, Layard R (eds) ¿Qué nos depara el futuro? Perspectivas desde las ciencias sociales, Alianza Editorial, Madrid, pp 87–138

Albornoz M, López Cerezo JA (2007) Presentación. Revista Iberoamericana de Ciencia, Tecnología y Sociedad (online) 8(3):43–46

Balaguer Bernaldo de Quiros C, Dormido Bencomo S (2007) Impactos sociales, económicos y laborales de la robótica en el área industrial y de servicios. In: Tezanos JF (ed) Los impactos sociales de la revolución científico-tecnológica. Sistema, Madrid, pp 449–494

Beck U (1992) Risk society: towards a new modernity. Sage Publications, London

Castells M (1996) La era de la información. Economía, sociedad y cultura. vol. 1. La sociedad red. Alianza Editorial, Madrid

Castells M, Hall P (1994) Las tecnópolis del mundo. La formación de los complejos industriales del siglo XXI. Alianza Editorial, Madrid

Collingridge D (1980) The social control of technology. Pinter, London

European Foundation for the Improvement of Living and Working Conditions (EFILWC) (2003) Handbook of knowledge society foresight. EFILWC, Dublin

Fowles J (1978) The handbook of futures research. Greenwood Press, Westport CT

Georghiou L, Harper JC, Keenan M, Miles I, Popper R (2008) The handbook of technology foresight. Concepts and practice. Edward Elgar Publishing, Cheltenham

Heidegger M (1975) Die Zeit des Weltbildes. In: Heidegger M (ed) Holzwege. Gesamtausgabe vol 5. Vittorio Klosterman, Frankfurt, pp 137–159

Hernández de la Fuente D (2008) Oráculos griegos. Alianza Editorial, Madrid

Hughes T (1983) Networks of power electrification in western society 1880–1930. The Johns Hopkins University Press, Baltimore

Kurzweil R (1999) La era de las máquinas espirituales. Cuando los ordenadores superen la mente humana. Planeta, Barcelona

Levy D (2007) Love and sex with robots: the evolution of human-robot relationships. Harper, London

López Peláez A (2003) Nuevas Tecnologías y sociedad actual: el impacto de la robótica. Instituto Nacional de Seguridad e Higiene en el Trabajo, Madrid

López Peláez A (2009) Prospectiva y cambio social: ¿cómo orientar las políticas de investigación y desarrollo en las sociedades tecnológicas avanzadas? Arbor. Pensamiento, ciencia y cultura 738:825–836

López Peláez A, Díaz Martínez JA (2007) Science, technology and democracy: perspectives about the complex relation between the scientific community, the scientific journalist and public opinion. Soc Epistemol 21(1):55–68

López Peláez A, Krux M (2003) New technologies and new migrations: strategies to enhance social cohesion in tomorrow's Europe. IPTS Rep 80:11–17

López Peláez A, Segado Sánchez-Cabezudo S, Kyriakou D (2012) Railway transport liberalization in the European Union: freight, labour and health towards the year 2020 in Spain. Technol Forecast Soc Chang 79:1388–1398

Morato A, Rodríguez A, Miles M, Keenan M, Clar G, Svanfeldt C (2002) Guía práctica de prospectiva regional en España. Directorate-General for Research and Innovation (EU), Luxemburg

Roco MC, Bainbridge WS (2002) Converging technologies for improving human performance. nanotechnology, biotechnology, information technology and cognitive science. National Science Foundation, Arlington (Virginia)

Ross A (1996) Science wars. Duke University Press, Durham

Schwartz P (2003) El río y la bola de billar: la historia, la innovación y el futuro. In: Cooper RN, Layed R (eds) ¿Qué nos depara el futuro? Perspectivas desde las ciencias sociales. Alianza Editorial, Madrid, pp 27–38

Smith MR, Marx L (eds) (1996) Historia y determinismo tecnológico. Alianza Editorial, Madrid

Technological Forecasting and Social Change (2006) 73(2):95–127

Winner L (1987) La ballena y el reactor: una búsqueda de los límites en la era de la alta tecnología. Gedisa, Barcelona

Chapter 3
The Impact of New Technologies in Organisations

Áurea Anguera de Sojo

3.1 ICTs, Environment and Organisations

In the last 30 years, the organisations have had to deal with many changes. Many of these changes stem from Information and Communication Technologies (ICTs).

ICTs have brought new ways of relating to the environment for organisations. Business prospects and expand business grow up with these technologies, especially the Internet and electronic commerce. New technologies have had a very significant influence on the environment in which organisations carry on their activity and on the way they relate to that environment. The influence of ICTs has been felt, to a greater or lesser extent, in all sectors of industry and of the market.

3.1.1 ICTs and Costs

The introduction of new technologies into industry involves changes in the manner of organising productive processes and, to be exact, it involves in general terms a lowering of costs as it makes possible organisational forms with a lower cost structure.

In this regard, the so-called transaction costs, put forward by the Nobel Economics Prize winner, Coase (1937), which are still absolutely valid in the digital world, deserve special mention.

Coase stated that companies were set up as a result of so-called transaction costs; his theory was that individuals come together in companies so as to be able to deal with what Coase called transaction costs as they cannot do so individually. Transaction costs are those which arise whenever there is a market operation or intervention and the following can be identified:

Á. Anguera de Sojo (✉)
Informatics and Law Department, Universidad Politécnica de Madrid (UPM), Madrid, Spain
e-mail: aureamaria.angueradesojo@upm.es

A. López Peláez (ed.), *The Robotics Divide*,
DOI: 10.1007/978-1-4471-5358-0_3, © Springer-Verlag London 2014

- Search costs: they are those that are derived from searching for suppliers of a certain product or service.
- Information costs: once the supplier is located, it is necessary to gather information on its solvency, history with other customers, previous transactions in which it has been involved.
- Negotiation costs, deriving from the determination of the terms on which the transaction will take place.
- Decision costs, related to the process of evaluation of the different offers to which access has been gained and a comparison between them.
- Policing costs, referring to the monitoring that has to be carried out so that the agreement is carried out on the agreed terms, and enforcement costs, when situations of breach arise.

The transaction costs set out by Coase continue to exist today. But ICTs contribute to reducing them as they supply tools which make it possible to carry out the activities they refer to with less effort and in a much shorter time, which leads to a substantial saving in the cost to any organisation of carrying them out; especially the Internet gives one a particularly suitable environment to achieve this reduction in transaction costs as the costs related to searching, information and negotiation become much cheaper using new technologies and especially the Internet.

Another of the great effects of ICTs is that they especially affect the sectors in which the treatment of information is of key importance. New technologies supply organisations and markets with the capacity to manage a huge amount of information and tools which help to allow that information to be used both inside and outside the company. The incorporation of information systems (IS) within organisations involves an advantage when managing information. Each organisation must decide what innovations related to the IS that it will incorporate and how it will do so. If they are capable of adopting those innovations into the key processes of the business, they will be able to generate greater opportunities for the organisation; but the scope that they decide to give to these innovations is a decision for each organisation, related to its attitude and its mentality with regard to new technologies. It is for this reason that organisations with the same information technology infrastructure obtain fundamentally different performances according to the use that they make of the technology.

Apart from the above-mentioned transaction costs and the costs related to obtaining and managing information, ICTs and especially the Internet contribute to reducing advertising costs—it is a much cheaper and more accessible medium than traditional methods of advertising, costs of purchases and sales—permitting access to new suppliers and customers, under more favourable conditions in which it is easier to compare different offers—and costs directly related to internal processes—such as those related to the management of inventories or human resources.

Information is key not only for organisations but also for customers/users who have access to a much greater quantity of information, making the market much more transparent to them. Over the Internet, they have access to information on companies, products, prices, services, opinions on the company or the products of other persons, etc., in a much more convenient and easier way than before the appearance of the Internet.

As has been explained in foregoing paragraphs, new information and communications technologies have led, for all sectors and in a generalised manner, to a significant reduction in costs.

3.1.2 How Organisations Adapt to ICTs

For many organisations, ICTs and especially the Internet have opened a window onto the world and they are now capable of reaching environments that it was formerly impossible for them to reach. But all of this has brought with it the need to adapt to changes and to bring them into the organisation.

The Internet is a market which can still be considered to be young. However, the expansion over the last 20 years has been spectacular. The figures make this clear. It is currently said that there are over 2.4 billion Internet users, or one-third of the world's population, according to the data from Internet World Stats (2012). Each one of them is a potential customer who uses the net to purchase products or services, to search for information or share experiences or knowledge using social networks, which have become basic communication tools.

The introduction of ICTs into the world of organisations and into markets involves the need to adapt to these new technologies. To be exact, the Internet implies, in the first place, being able to reach market segments to which many companies did not have access for purely geographical reasons. The Internet increases the radius of action of companies and of organisations in general. In the second place, adaptation to these technologies has led to change in certain internal structures of the company which have had to adapt to deal with the new circumstances.

The organisations are conscious of the power of the Internet and of ICTs. It is not just a matter of doing business by selling products or services but of being present on the net and reaching persons/customers that a few years ago it was impossible to reach. Because it is also true that not being present on the net means, according to some people, "digital death". The digital world has brought about a change of mentality in the traditional ideas of doing business. This change is directly linked to the attitude that each organisation—or rather, the leaders of the organisation—adopts to innovation, involving the adoption of new technologies at the heart of the organisation, and new formulas to gain new customers and keep their loyalty. ICTs are advancing very fast and this means that it is necessary to develop a great capacity for adaptation and reaction to these changes.

The different systems of the organisations have to be modified to respond to the new environment. If in the previous economy, in most cases, the external part of the company was hardly related to the internal part, in the digital era they are both very closely related. The information flows between the two parts have become much more intense and a greater volume of information is handled. This means that the new IS which support the company have to be able to respond to this new situation and also to guarantee that the information is accessible to whoever needs to have access at the moment at which he must have access. Therefore, organisations have been required to introduce IS to which, as was commented in the foregoing section, each one has given a different scope.

New companies and businesses have arisen backed by the Internet—the dot-coms—and new kinds of relationships between them. Both those organisations that were already established before the Internet and the dot-coms have had to reposition themselves within the sector in which they carry out their activity and thereby they have been modifying the structure of the markets.

On the one hand, environments such as that supplied by the Internet provide almost immediate access to information on customers, suppliers, competitors and above all what affects the market in general. As has been indicated above, it is possible to contact other participants in the market, which was formerly not possible as there were geographical barriers.

These new connections or relationships make the balance of power in the market change and mean that the influences that are derived for the organisation from these balances also undergo important variations.

Analysing the existing literature on the impact that ICTs have had on the market and in organisations, two well-differentiated lines appear which are opposed in their reading of how the new situation should be faced.

3.1.2.1 Downes and Mui's Radical Change

For authors such as Larry Downes and Chunka Mui, the onslaught of new technologies in the economy has led, above all at the initial moment, to confusion in organisations. The main problem that companies come across is that of attempting to deal with ICTs with the same postulates that they had been using up to this time.

Downes and Mui (1998) understand that the postulates of the traditional economy, fundamentally the principles listed by Porter on competitive strategy, have no place in the digital era. The guidelines of the new economy and the tools supplied by the new technologies cannot follow these principles. The changing and, to a large degree unpredictable, environment in which organisations are immersed is not compatible with traditional postulates, specifically with strategic planning, of which Porter is the greatest defender.

If planning was formerly undertaken for 5 years, now organisations have to make plans for 18 months and prepare responses much more quickly than in the past. There is greater access to information to back up decisions, but they must be taken much more quickly. For this reason, the decision-making processes have changed substantially and with them the structure of the organisation and the way it acts.

We find ourselves with the need for a radical change within organisations so as to be able to adapt to the new environment and make use of the advantages that it offers. Organisations cannot act in the new environment with the same parameters that they had defined for a more static environment. Because, if anything characterises the new environment, it is dynamism.

On the other hand, the world of new technologies is evolving in a vertiginous manner. On many occasions, the technology which is new today will tomorrow have ceased to be new because something newer and better will have arisen. Technological changes behave exponentially, changing very rapidly in a very short time.

In this changing, unpredictable and extremely dynamic environment, what Downes and Mui call killer applications arise, which are inventions which apparently introduce totally new categories, in such a manner that, as they are the first, they dominate that category and produce incredibly high returns on investment. Killer applications do not constitute a new concept in themselves as throughout history there have always been innovations which have meant a transcendental change in their category with regard to how things were done before; let us think for example of the appearance of the pulley or the steam engine, which were decisive moments, each in its own sector. The difference with the previous examples is that in the technological sphere these inventions arise much more rapidly and are propagated much faster. And the Internet is especially important to this propagation as it disseminates these innovations and uses them to get better and better.

In this context, the managers of organisations perceive the technology as competition and not as a tool to help them. The perception that they have is that if they are not capable of accepting and introducing those changes within their organisations, they are outside the market and are being left behind by it.

In the opinion of Downes and Mui and their disciples, in order to prevent this situation, it is necessary to change strategies, leaving aside the principles of the traditional economy and implanting a digital strategy which is indeed capable of collecting and using for the benefit of the organisation all these new developments. Killer applications impose a different speed, in such a manner that the activities which are capable of adapting to them survive and those which are not disappear.

The digital strategy which they propose for organisations is based on accepting the new situation and applying different variables to those which were applied in the traditional market. There are new forces involved in the market and organisations must understand and adapt to the new market structure. The new forces are digitalisation—communications increase exponentially and significantly reduce the costs—globalisation—the world is a great network and is a great market for organisations—and deregulation or liberalisation—as it is a market which is not controlled and in which it is the purchasers and sellers who regulate it by their actions. These new forces determine a different kind of strategy, in the opinion of some, to the traditional competitive strategy which was the basis of the traditional market.

This digital strategy is based, not only on generating a sustainable competitive advantage but also on the capacity of organisations to generate new connections, reshaping the environment—in the sense that the question must constantly be asked internally which operations they must continue to maintain and which must be outsourced—and redesign the interior—as organisations in a certain manner have to reinvent themselves in the digital world, their internal structure must evolve in a digital world.

3.1.2.2 The Tradition of Porter

Against the posture defined in the foregoing section, there are authors who understand that ICTs do not involve a clean break with all the past and that the job of organisations in this new environment is to adapt to it without losing the values inherited from what we might call the traditional economy.

However, it is necessary to indicate that the adaptation to this change has not, in many cases, been carried out in the proper manner. As Porter (2001) states, the initial fervour that the Internet gave rise to led some companies, and especially the dot-coms, to take wrong decisions. And they did so because they were working from false premises. The arrival of the Internet has led to euphoria in companies and in markets which understand the growth opportunity that they represent but which falsely believe that the Internet involves a break with the past, with the traditional organisational and competitive rules; and this is not so.

If the success of an organisation was based on finding a competitive advantage (something in which they are better than others in their environment) and making that advantage sustainable over time, in the early days of the ICTs, it was believed that this was no longer necessary. There has been a confusion, in the opinion of some writers such as Porter, between the tool (the ICTs) and the competitive advantage, and it has been assumed that the adoption of ICTs within an organisation constitutes its competitive advantage.

Therefore, with the arrival of the ICTs, many organisations reacted by abandoning what had up till then been their strategy and centred their business on the adoption of these new technologies. And this can lead the organisation to lose its positioning in the market as it sets aside what had made it different from the others up till then.

As regards dot-coms, the problem that the majority of those which arose initially had is that they did not have a business plan to back them up. They are organisations which arise, creating great expectations and attracting a lot of capital but with nothing solid behind them. For this reason, once the initial boom of the dot-coms had passed, a significant percentage disappeared.

In the opinion of Porter and his defenders, the so-called digital era does not represent a total break with the past. It is true that new forces appear in the market which organisations must take into account—digitalisation, globalisation and deregulation—but this does not invalidate the five forces described by Porter which continue to be valid, although affected by new technologies and especially the Internet.

As has been indicated above, the positions that suppliers and customers may have within a market—two of the forces pointed out by Porter—undergo modifications through the appearance of other forces such as globalisation or digitalisation, which have modified the influence of each of them on organisations participating in the marketplace. Thus, in general, suppliers have lost power and customers have gained through the generalised use of the Internet. The entry barriers have also been affected as in the new situation they have lost importance in the context of the environment.

And it is for this reason that the competitive advantage, in Porter's opinion, has become more necessary with the appearance of the Internet due, among other things, to the changes in the internal power balances of the markets.

As Porter indicates (Porter 1996), there are six fundamental principles so that a company can establish and maintain a competitive advantage:

1. It must start with the right goal: superior long-term return on investment.
2. A company's strategy must enable it to deliver a value proposition, or set of benefits, different from those that competitors offer.
3. Strategy needs to be reflected in a distinctive value chain. A company must perform different activities than rivals or perform similar activities in different way.
4. Robust strategies involve trade-offs. Las compañías deben centrarse en aquellos servicios, productos o actividades en los que son mejores que la competencia, y esto implica abandonar aquellos otros en los que no los son. Por tanto, la estrategia puede implicar sacrificios para la compañía que ésta debe estar dispuesta a asumir en aras de mantener su ventaja competitiva.
5. Strategy defines how all the elements of what a company does fit together.
6. Finally, strategy involves continuity of direction. A company must define a distinctive value proposition that it will stand for.

The principles indicated have been criticised over the last few years as there are authors who understand that they have no place in the so-called digital economy. To be exact, Porter's detractors understand that in an economy in which the changes are produced in a very rapid and occasionally sudden way, conceiving the competitive strategy as something which must last over time is in contradiction to the economy itself. Because companies must be flexible and adapt or change their strategies according to the changes which arise in the environment. And if it is not working, it is better to change.

Even so, the competitive strategy must be the basis for any organisation, now as much as before. Because what determines success is the capacity to generate that something which is different, which distinguishes the company from others and which means that in a global marketplace, such as the Internet, the company is differentiated from all the others that compete with it. The problem, which is also derived from the Internet, is that in this medium, it is much more difficult to achieve that competitive advantage but it is also true that for the same reason, once it is achieved, it will be much more difficult for competitors to imitate.

In order to achieve this strategy, the behaviour of the organisation must be directed towards integrating new technologies into its strategic thinking and achieving the best possible performance from them; but this is not the same thing as making ICTs the strategy. They are a tool, not an end in themselves.

On this line of thought, there are some people (Murad 2001) who propose a return to the traditional, in the sense that after the initial "hysteria" caused by the Internet, it is necessary to recover the basic parts of the past and build the future integrating both things and not suddenly breaking with the past and considering that nothing from before the times of the Internet is valid today.

3.2 Management and ICTs

Over the last 25 years, new information and communications technologies, as has been indicated above, have entered the organisational sphere, leading the managers to have to face up to the need to introduce them into their organisations and use them to manage to reduce their costs and to compete and to survive in the new channel that the Internet has brought about.

Traditional and digital companies use the Internet and the communications networks to transmit information, manage it and also to establish electronic links or connections with their customers, suppliers or with other organisations.

The management of all this information is carried out using IS which acquire a significant role within the organisations due to the quantity of information that they manage and because they are the basis which supports the taking of the organisation's decisions. In a very dynamic and unpredictable environment, the possibility of reducing the uncertainty in the taking of decisions is a fundamental parameter for the management of an organisation.

But also, the IS allow the organisation to establish relationships with customers, suppliers, other companies, and to manage the activities or exchanges that they undertake with them. And it must not be forgotten that all these activities are undertaken, more and more, over the Internet.

3.2.1 Information Systems

An information system can be defined (Laudon and Laudon 2004) as a set of interrelated components which recover, process, store and distribute information to support the taking of decisions and the control of an organisation. Understanding information to mean any datum which has been treated in any way which is useful to the human being.

IS carry out three basic activities: collecting information, processing it and making it into output, so that it can fulfil its purpose. For this reason, the information must be received by the people who are responsible for taking the decision that it supports.

The activities carried out by the IS give a perspective of the elements that they are part of. On the one hand, the equipment (hardware) which stores the information and in which the data are processed to turn them into information; the software which constitutes the logic for the processing of the data, which must be defined according to the particular needs of each organisation; and finally, the persons who make up and give support to the system and those who must use the information which this supplies (users of the system).

IS provide value to the organisation, given that they support the taking of decisions and also improve the execution of the business processes which has a positive effect on the value of the organisation.

The IS are having greater and greater importance for the organisation; as they enter into an increasing number of departments of the company and pass the information on from one area to another, both internally and externally, they become a key element in the administration of the organisation. This means that any change in the IS has an effect on the processes, procedures or strategies that they support and vice versa, any change which is made to one of those processes, procedures or strategies must be properly taken up by the information system which supports it.

For an information system to be able to fulfil its function in the organisation, it must be capable of receiving and processing data without errors and generating useful information for the system users. Furthermore, it has to be able to guarantee the integrity and security of the information as well as the proper availability in time of the information.

There are different classifications of IS. Perhaps the most useful is that proposed by Laudon and Laudon (2004) in which IS are classified according to their usefulness at the different levels of the organisation—operational level, knowledge level, administrative level and strategic level:

1. Operations processing system: they are responsible for the administration of those daily routine operations that are necessary in business management such as applications of payrolls or monitoring of orders. These systems generate information which will be used by the remainder of IS in the company and will be used by the staff at the lower levels of the organisation (operational level).
2. Knowledge working systems: those IS that are responsible for supporting the agents who manage information in the creation and integration of new knowledge for the company; they form part of the level of knowledge.
3. Office automation systems: systems used to increase the productivity of the employees who manage the information at the lower levels of the organisation (word processor, electronic diaries, spreadsheets, e-mail, etc.); they are classified at the level of knowledge.
4. Administration information systems: IS at the administrative level used in the process of planning, control and decision making supplying reports about ordinary activities (inventory control, annual budgets, analyses of investment and financing decisions). They are used by the management at intermediate levels of the organisation.
5. Systems for supporting decisions: interactive IS which help the different users in the process of taking decisions, when using different data and models for the solution of non-structured problems (cost analysis, price and profit analysis,

analysis by sales per geographical area). They are used by the intermediate management of the organisation.

6. Systems of management support: IS at the strategic level of the organisation designed for taking strategic decisions by means of the use of graphs and advanced communications. They are used by the senior management of the organisation with the aim of preparing the general strategy of the company (planning of sales for 4 years, operational plan, labour planning).

The evolution of the IS over the last few years has led them from a first stage at which their main function was to mechanise and automate the ordinary processes of the organisation at the current stage at which it is possible to talk of IS strategies in which the system itself is integrated in the strategy of the organisation, passing through the integration phase of the different IS of the organisation.

In order to reach the final phase, the persons responsible for the organisation have to be conscious of the value of the integration of IS in the organisational strategy or of considering it as one more strategic element. The development and the profits that IS can give to the organisation will depend on this consideration.

Over the last 10–15 years, certain IS which concentrate on two fundamental aspects and which have been used within the strategy of adaptation to new models of the economy in the sphere of new technologies, and especially the Internet, have undergone special development in organisations.

3.2.1.1 Enterprise Resource Planning

Enterprise resource planning (ERP) systems are management IS which handle many of the pieces of business associated with the operations of production and the aspects of distribution of a company in the production of goods or services. In other uses, they serve to automate many of the business practices associated with the operational or productive aspects of a company.

ERPs handle information related to the production, logistics, distribution, inventory, despatches, invoices and accounting of the organisation. However, this is a system which can intervene or act with many other business activities such as sales, deliveries, payments, production, administration of inventories, quality of the administration and the administration of human resources. They are management systems for the company.

They are habitually made up of different modules. These parts have different uses, such as production, sales, purchases, logistics, accounting (of various kinds), project management, geographical IS (GIS), inventories and warehouse control, orders, payrolls. Apart from supplying updated information on each of the matters covered by the different modules, their main objective is to give support to the customers of the business, making it possible for the response times to their problems to be as fast as possible, as well as an efficient handling of information which makes it possible to take decisions in a timely fashion with a reduction in the total costs of the operation.

ERPs are modular, configurable systems as they have to adapt to the different organisations. The personalisation and private development for each organisation may on occasions be a tough task which requires a lot of time so as to be able to model all the business processes which it must cover. And it therefore often has a high cost.

Among the limitations and obstacles of ERPs, the following ones can be pointed out:

- The success of the implementation of the system depends to a great extent on the abilities and experience of the workforce, including their education, and knowledge of how to ensure that the system works correctly.
- The installation of an ERP system is very costly.
- ERPs are perceived as very rigid systems which are difficult to adapt to the specific flow of workers and the business process of some organisations.
- Sometimes these systems can be difficult to use.
- The systems can suffer from bottlenecks, which means that the inefficiency of one of the departments or one of the employees can affect other participants.
- Resistance to sharing internal information between departments can reduce the efficiency of the software.

3.2.1.2 Customer Relationship Management

Customer relationship management (CRM) arose as a response to the need of organisations to maintain close contact with their customers. It can be considered to be the technological solution to be able to develop so-called relationship marketing, understood as the process of identifying, creating, satisfying, retaining and strengthening (and when necessary terminating) profitable relationships with the best customers and other groups, in such a manner that the objectives of the parties involved are achieved (Grönroos 1997). These relationships should be, in the long term, interactive and should generate added value.

CRM strategies have as their objective, on the one hand, to reduce costs and, on the other hand, increase income. They involve total focus on the customer, in which each customer is treated in a unique manner, which is very much in consonance with the guidelines for the digital economy postulated by Downes and Mui, which have been analysed above.

CRM makes it possible to measure and monitor each interaction between the customer and the organisation. What is more, it is possible to know the result of these interactions and to calculate the return on each of the efforts that have been made. With this information, it is possible to get to know the profitability per customer or group of customers and, as a result, decide on the amount of resources to use to give them services based on the said profitability. It is possible to have the relevant information on prospecting (decisive contacts, activities carried out, proposals presented) and preferences on products/services. All this information enables the organisation to have an effect on those customers or products/services which are most profitable for it.

CRM also makes it possible to hold on to customers which may be a basis for the future throughout the company. Managing to have satisfied customers reduces the advertising costs and, on the other hand, makes customers less sensitive to offers which might reach them from competitors, which guarantees their loyalty to the organisation to a certain extent.

3.2.2 The Organisation on the Internet

If the new information and communications technologies have meant a revolution for the world of organisations, it is no less true that the technology which has had the most impact on the future of organisations has been, and is, the Internet.

As is explained in Reina (2011), the application of the Internet is not limited to the IS of the company but opens up the company to its relationship with the environment and integrates these systems in the flows of exchange of services and information with that environment. The way of relating to that environment is the web system of the company and the design of the same depends on the fit that is desired in the infrastructure of the organisation and in its strategic planning.

Kotler et al. (2000), the web environment has supplied direct marketing tools to the organisation as well as relationship marketing and has opened up new routes to relationships which the company might establish with its customers.

The actions of organisations in this sphere began with strategies for positioning in the search engines and have now reached social media optimisation (SMO) strategies.

SMO strategies are linked to the use of social networks and of virtual communications with a commercial or advertising purpose. There are more and more organisations which have their profile on networks such as Facebook or Linkedin. The tools to develop this kind of strategy are highly varied and grow day by day with the appearance of new tools. Among others, the following should be mentioned:

- Blogs: a periodically updated site which chronologically compiles content from one or various authors.
- Microblogging: this is a service which permits users to send and publish short messages.
- Sites with shared content, such as Flick or YouTube.
- Wikis, of which the leading one is Wikipedia.
- Social networks, whether horizontal (Facebook) or vertical (Linkedin).

The value of integrating traditional and Internet techniques allows organisations to create advantages by deploying Internet technology to reconfigure traditional activities or finding new combinations of Internet and traditional approaches.

3.3 Human Resources and New Technologies

The arrival of ICTs in organisations has led to some changes in the structure of their human resources. On the one hand, it seems to be necessary to redefine the role of the manager or senior manager in the organisation, who must adapt to the new environment and ensure that the organisation also does so. And on the other hand, the human resources, the human capital of the organisation is affected in different ways by the entry of the tools related to new technologies.

Human resources and their management serve as a strategic asset to the organisation. There is a scholarly debate about what is specially the decision factor, human resources or the way they are managed.

Definitions of human resources understand them in two ways; the first one (Fisher 2006) defines human resources as people who works in/for an organisation. Second approach (Kamoche 1999) takes into consideration not only people but also their skills, knowledge, abilities, attitudes and experience. And it has highlighted the importance of human resources in the enhancement of organisational effectiveness and competitiveness.

Kamoche (1999) maintains that, on the one hand, human resources are essential to the development of competitive advantage, and, on the other hand, human resources management (HRM) leads to the possession of organisational ability to align human resources with strategy, as well as retaining such human resources.

Human resources play a critical role in the enhancement organisational strategy and competitiveness. Therefore, HRM should be considered as a strategic activity and, therefore, should be carried out in a completely consistent way with the overall organisation strategy.

3.3.1 ICTs and Management

As noted above, the HRM is a critical factor in the deployment of the competitive organisational assets.

On how ICTs have affected the figure of the senior manager, there is a division of opinions. As has already occurred in the matter of understanding how these technologies affect organisations, in the case of the manager, there is a sector with the opinion that the bosses in traditional organisations cannot act in this new environment, while others understand that although it is true that the Internet has changed the abilities required in businesspeople, it has not done so in a drastic manner. That is to say that if a person is a good manager in a traditional company, he may also be a good manager in an Internet environment.

It is true that the abilities have changed but not the management or at least not in such a decisive manner that managers cannot be retrained to develop those abilities which had not been so necessary up to the present.

The ten points which the e-manager must bear in mind are (The Economist 2000):

1. Speed: technology contributes to reducing the duration of production cycles. It is necessary to act in a dynamic manner so as not to lose access to the market. As it may not be possible to carry out all the activities in a simultaneous manner, it is necessary to establish priorities.
2. Good Staff: part of the adaptation of the organisation to the new times involves adjusting the jobs and roles of the people who occupy them, in such a manner as to get each one of them to contribute the maximum value to the company.
3. Openness: this is a key point of the Internet strategy but it is necessary to be careful with the degree of exposure that may be given of the weaknesses of the organisation towards the exterior.
4. Collaborative abilities: it is indispensable to develop communicative and team working abilities as the Internet is a supremely collaborative environment.
5. Discipline: in order to avoid the organisational chaos that entering the Internet can sometimes lead to for organisations.
6. Good communications: above all internally, to explain the new processes and avoid misunderstandings.
7. Content management abilities: it is necessary to have absolute control of what is communicated to the exterior. The handling of information is an advantage for those organisations that manage it.
8. Focus on the customer: via the methods of relationship marketing which have been commented on above.
9. Administration of knowledge: the databases and intranets help with this task. It is not easy to bring the brilliant ideas of employees to the attention of the management but an effort has to be made to do so.
10. Leadership by example: the first people who have to be involved in the use of ICTs should be the managers so as to give an example with their actions.

3.3.2 ICTs and Human Resources

In the last years, there have been substantial transformations in the way organisations perform their activities. The use ICTs has modified numerous functional areas and has changed the work environment of the organisation. It is also realised that in alignment with the impact of ICTs, there have been placed another changes in organisations related to them.

Some authors maintain that the competitive advantage lies in the human resources. To build an organisational capability, employees competences need to be developed and also retained through effective measures (Boxall 2003).

Scholarly literature is plenty of different investigations about relationships between technological changes and organisational changes in human resources.

These investigations have found some important findings (Bayo-Moriones et al. 2008):

1. Positive correlation between ICTs and more skilled workers. In the field of human resources, in the last two decades, many countries have registered changes in the skill composition of labour as well as a trend of upskilling. Computerisation and ICTs increase production data which require workers to show analytical abilities in order to process the information. ICT also raises the amount of knowledge available for employees; to obtain the maximum benefit from this knowledge, there are needed higher skill employees to assimilate, integrate and communicate it within the organisation in a more efficient way.
2. Due to the increase in the rate of ICT adoption and use within organisations, advance economies have shown a trend of upskilling and a remarkable variation in the skilled/unskilled composition labour.
3. New skills and abilities are required in this new scene. Data analysis capacity, reasoning and problem-solving abilities and communication skills are some of the competences which organisations are searching for in their workers; and they are often found in higher education individuals.
4. Organisations invest in training. In this context, training activities play an active and leading role within organisation. As ICT tends to increase the level of required qualification, training becomes an important fact in order to improve workers skills and abilities. ICTs contribute to this organisation aim with new ways of learning, such as e-learning technologies, which provides contents and capacity to train human resources.
5. The effective use of ICT requires innovative work practices. ICT is associated with internal organisational changes. ICTs often entail a higher degree of authority decentralisation, which means decentralisation in decision-making process. Structures such teamwork or self-managing teams are used instead of other inflexible structures.

These are some of the most important impacts of the ICTs in human resources and their management. The relationship between new technologies and changes in human resource structures has been empirically demonstrated among several studies in last years. The effects of these changes will be showed in the next years. Perhaps there will be new ones or some of them tend to disappear, but for the time being these above are the most important ones. Human resource management should be viewed as a strategic activity and thus carried out with the overall organisation strategy.

References

Bayo-Moriones A, Billón M, Ler-López F (2008) Skills, technologies and organisational innovation in Spanish firms. Int J Manpower 29(2):122–145 (Emerald Group Publishing, Bradford)

Boxall P (2003) Strategy and human resource management. Palgrave Macmillan, Houndmills

Coase RH (1937) The Nature of the firm. Economica 4:386–405

Downes L, Mui C (1998) Unleashing the Killer App: digital strategies for market dominance. Harvard Business School Press, Boston

Fisher CD (2006) Human resources management, 6th edn. Houghton Mifflin Company, Boston

Grönroos C (1997) Value-driven relational marketing: from products to resources and competencies. J Mark Manage 13:407–420

Internet World Stats (2012) http://www.internetworldstats.com. 30 de junio de 2012

Kamoche K (1999) Strategic human resource management within a resource-capability view of the firm. In: Jackson RS (eds) Strategic human resource management. Blackwell Publishers Ltd, Oxford

Kotler P, Cámara D, Grande L, Cruz L (2000) Dirección de marketing. Edición del milenio. Décima edición. Pearson Educación, SA, Madrid

Laudon KY, Laudon J (2004) Sistemas de información gerencial. Octava edición. Pearson Educacion, Mexico. 970-26-0528-8

Murad DS (2001) Back to fundamentals. The ChemQuest Group, Inc

Porter M (2001) Strategy and the internet. Harvard Bus Rev, Mar, pp 63–78

Porter M (1996) What is strategy? Harvard Bus Rev, Nov/Dec, pp 61–78

Reina M (2011) Marketing. Pasado, presente y future. Sanz y Torres, Madrid

Inside the Machine (2000) A survey of E-management. In special report. Economist

Chapter 4
Economic Impact of Smart Specialization and Research in Advanced Adaptive Systems in a Monetary Union

Dimitrios Kyriakou

4.1 Introduction

Tensions in the eurozone have brought back to the surface issues/criteria related to its solidity and optimality—the old optimum currency area debate. Research has drawn comparisons between intra-eurozone heterogeneity and divergence, on the one hand, and the heterogeneity and divergence between US Federal Reserve System regions. Higher divergence in Europe has made central bank policymaking more difficult and has not been as successful and as agile in dealing with the crisis as the US Federal Reserve has been (Sheets and Sockin 2013). We will suggest that technology and research can help buttress a monetary union, through their impact on trade and specialization, and production structures.

The current economic crisis and more recently the sovereign debt crisis have increased pressure on countries to redress structural problems in their economies, while preserving the margin for public investment in knowledge-based capital, in a "smart way," contributing to productivity growth and competitiveness. Smart specialization, the concept and the associated policy process, aims to do just that, fostering growth potential in a context of rapid technological change and globalization. The rationale for smart specialization stresses the role of policymakers, knowledge-based institutions, and entrepreneurs in shaping specialization and competitiveness. Horizontal key enabling technologies play a particularly essential role in boosting existing strengths, as well as revealing new economic opportunities in sectors at various levels of technological sophistication.

This applies particularly to advanced adaptive systems, such as robotic systems that can be used in wide arrays of sectors to enhance performance and competitiveness. This will help European economies to compete internationally by emphasizing their strong card, technology, as opposed to competing on labor cost, on which they could hardly hope to win. Moreover, and more to the point here,

D. Kyriakou (✉)
Institute for Prospective Technological Studies, Seville, Spain
e-mail: dim.kyr6@gmail.com

A. López Peláez (ed.), *The Robotics Divide*,
DOI: 10.1007/978-1-4471-5358-0_4, © Springer-Verlag London 2014

it will help mitigate the tensions caused by divergent performances between members of a monetary union, through promoting convergence toward higher technological plateaus and technoeconomic convergence-driven intra-industry trade (henceforth intra-industry trade for short).

This emphasis on technoeconomic convergence and intra-industry trade has a counterpart, in terms of technology and innovation policies, as one needs to foster wide spectrums of industries in many countries, which can hardly compete globally on the basis of labor costs. The only way to pull it off is to promote technology- and innovation-driven competitive advantages, exploiting the existence or development of varieties that are imperfect substitutes between them. Moreover, the technological drive provides for continuously improving varieties, keeping ahead of the competition, generating wider arrays of varieties, and nurturing consumer interest.

As a first step, a currency area consists of a group of countries featuring either a common currency (full monetary union) or permanently fixed exchange rates among their national currencies which are furthermore convertible into one another. The exchange rates of these currencies with currencies not participating in the currency area are flexible (note, however, that the exchange rate between a member of the group and an outsider is indirectly dictated by previously set exchange rates between other members of the group and the particular outsider).

An *optimum* currency area on the other hand is not easy to define. The problem stems from the tendency of many writers to define as optimum those areas that pass their particular test for optimality. Since several tests and criteria have been suggested, the practicality of such an approach is dubious. It is therefore best to define an optimum currency area (henceforth OCA) by its desirable characteristics and not by the minimal set of criteria which allegedly guarantee its optimality.

Machlup suggests (Mundell and Swoboda 1969) that in an OCA, member countries can simultaneously maintain both internal and external balances and adopt fixed exchange rates with the other members of the OCA and guarantee perfect convertibility. McKinnon is more specific (as quoted in Mundell and Swoboda 1969, p. 42) requiring that an OCA guarantee the maintenance of full employment, of balanced international payments, and of a stable internal average price level. According to McKinnon under flexible exchange rates, depreciation will increase the demand for "tradeable" output and will lead to firms luring away employees from the production of non-tradeables by offering higher wages which will generate inflationary pressures.

4.2 Proposed Criteria

In reviewing the conditions that have been suggested as sufficient for optimality, we will start with Mundell. He argued (Mundell and Swoboda 1969) that perfect factor mobility among member countries was a sufficient condition for the establishment of an OCA; effectively, he reduced the problem of adjustment to the inter-regional case within a country. For example, a drop in the demand for the products of a particular region will raise unemployment in that region; perfect capital mobility, however, will boost local demand and employment, with the help of capital flowing in from other regions in the country in search of lower wages. Furthermore, labor mobility allows unemployed workers to seek employment in other parts of the country.

Two caveats should be mentioned here which effectively make Mundell's criterion a necessary but not sufficient condition: First, what we need is not simply labor mobility but rather occupational mobility and a way to overcome language barriers. This implies either that we have to limit OCAs to microregions with all the concomitant problems (too many currencies, no store-of-value currency, no standard-of-value to allocate capital among regions for international investors, and high cost of hedging in forward markets which might be too thin in such a semi-anarchic situation) or that we need effective retraining and relief programs to help the region through its hardship—a role played by the government in a country but which economic models of OCA do not presuppose.

Second, disequilibria may persist if there is perfect factor mobility across countries, but not across all industries. Differing degrees in labor intensity in production among members may perpetuate high unemployment rates (e.g., the effects of a drop in demand on a labor-intensive industry such as construction may not be fully reversed if jobless workers can only move to capital-intensive industries such as chemical processing and if the capital required per product unit in the capital-intensive industry is sufficiently inelastic with respect to changes in the relative rewards to the factors of production). Furthermore, the capital flowing into the troubled area if not appropriately invested may exacerbate the situation by financing higher consumption of imports, thus leading to heavier loan repayment schedules in the future—this last point touches on the issue of "imaginative action" on the part of the government, a term employed by Ingram (Mundell and Swoboda 1969).

Let us keep in mind before proceeding to other criteria that the inter-regional or inter-sectoral problems alluded to above are not particular to fixed exchange rates. Under flexible exchange rates, the same problems are encountered. Flexible exchange rates clear the foreign exchange market, but not necessarily the domestic goods and factor markets.

McKinnon proposed (Gandolfo 1987) the degree of openness of the economy as a condition for OCA. The degree of openness is proportional to the ratio of the country's internationally traded products to the GDP; a country with a high ratio can benefit from participation in such an area. Such a country has a small non-tradeable goods sector which would have to undergo huge disturbances to accommodate shifts in prices and employment induced by exchange rate fluctuations (e.g., a depreciation to improve the current account). It would be better for such a country to adopt fixed exchange rates and pursue contractionary fiscal policies to reduce imports and improve the current account. Fixed exchange rates, however, cannot preclude the painful resource reallocation if the exchange rate fluctuation is due to fundamental changes. In that case, we encounter again the problems of occupational mobility, language barriers, persistence of disequilibria, mentioned above in treating Mundell's view. Furthermore, McKinnon's view taken to the extreme leads to a rather peculiar prescription: the foregoing of ER manipulation by a country which bases its economy on the revenue from the export of one product. Such a country will suffer severe disturbances if competitors appear in the world market or if foreigners' tastes start to change, and it cannot resort to depreciation measures.

Kenen, contrary to McKinnon, suggested that excluding large macroeconomic disturbances (such as general inflation), well-diversified economies would be less vulnerable to changes in their terms of trade, compared to highly specialized economies, since high- and low-performance exports will coexist side by side. He further argued that in such an economy, a fall in the demand for its exports will not generate a sharp rise in unemployment and that links between external and domestic demand (especially the link between exports and investment) will be weaker in well-diversified national economies. Objections have been voiced against the sufficiency of this criterion, too. Kenen himself points out in the same article the poorer performance of well-diversified economies when the exogenous shock involves wage rates increasing faster than import prices as well as diversification's inability to protect against instability imported from abroad, e.g., when faced with demand for exports dropping due to business cycles. Furthermore, he concedes to McKinnon that the likely small size of the foreign trade sector in a diversified economy undermines his argument and the size issue can lead to greater instability when monetary and fiscal policy tools are used to pursue internal balance (Kenen in Mundell and Swoboda 1969).

Another criterion, proposed by Scitovsky, is the degree of financial integration among candidate participants in a currency area which focuses on the role of capital flows in equilibrating payment imbalances (Hallwood and MacDonald 1986). A group of highly financially integrated countries can proceed to form an OCA. In this case, exchange rate manipulation is not necessary to correct external imbalances since small changes in interest rates can generate the desired equilibrating capital flows. Several objections can be presented against the sufficiency of this criterion.

To begin with, complete financial integration may indeed abolish the disturbances due to exchange rate fluctuations and thus make flexible exchange rates more attractive. Second, the small interest rate differentials necessary must be conceded upon by the other members of the OCA (again pointing to the need for a guiding authority). The burden of future interest payments—entered under "investment income" in the services account—leads to a deterioration of the current account. Eventually, increases in the interest rate in order to attract new capital may make the burden of interest payments exceed the inflow of new capital, making the balance of payment schedule backward sloping (meaning that beyond a certain point, an increase in the interest rate will have to be accompanied by a currency depreciation to preserve equilibrium in the balance of payments). Such increases may exacerbate the balance of payments and call for lower national income growth in order to reduce imports and/or efforts to reduce the marginal propensity to import (e.g., through advertising "buy USA" or "boycott S. Korea"). Finally, even if interest is not repatriated, viewing these capital flows as not pure flows but as stock adjustments, diversification-of-portfolio principles put an upper bound on the capital that will flow into a country for a given interest differential (Gandolfo 1987).

Yet another criterion that has been proposed is the similarity in rates of inflation since divergence in inflation rates leads to changes in the real ER, whereas it is alleged that there will be no such effect if inflation rates are identical—or similar. Although this also is a necessary condition, it is far from sufficient. In several cases, in recent history, economies have had similar rates of inflation without satisfying other criteria for OCA (e.g., Grmonetary unione, FRG, Finland, and Canada in the 1960s and early 1970s). Furthermore, one of the disappointments of the experience with floating exchange rates (e.g., in the 1970s) was that changes in nominal exchange rates did not track purchasing power parity (PPP) and indeed affected the real exchange rate (MacDonald 1988).

The last criterion is almost a tautology since it calls for policy integration ranging from coordination of economic policies to surrendering monetary and fiscal sovereignty to supranational authorities. Clearly, this implies some degree of political integration.

4.3 Costs and Benefits Associated with Currency Areas

Besides the traditional approach of trying to provide a single sufficient condition for OCA, an alternative approach focuses on the costs and benefits incurred by a country contemplating membership in a currency area. The main benefits that have been suggested are numerous. First, a common currency can act as the standard and store of value as well as eliminate hedging costs on the forward market. It can improve allocative efficiency, integration of production, and economic specialization. Second, in a currency area, speculations on exchange rates (ER) would be eliminated, if the ERs are credibly fixed. Otherwise, destabilizing speculation such as in the Bretton Woods period could take place. Third, members do not need international reserves since transactions within the area can be completed using local currency. In the beginning stages of the monetary integration, however, it will be necessary to hold reserves in order to ensure ER rigidity and make this rigidity credible in the future so that the advantages mentioned above can be realized. Fourth, risk pooling will occur as economic misfortune in one area may be offset by fortunate developments in other areas, and the flow of foreign exchange reserves for the area as a whole would remain stable. Fifth, monetary integration may stimulate economic policy integration, and a commitment to maintain fixed exchange rates will make anti-inflationary policies credible and will impose discipline.

Note, however, that the discipline arguments common also to fixed exchange rates and gold standard systems have not been corroborated by historical developments. Also, note a circularity in this argument which renders inflation convergence on outcome of the integration process, whereas it was previously deemed a prerequisite of it.

Benefits may be reaped (seigniorage due to the increased use of the common currency as international reserve currency) as a result of the larger weight carried by the currency area than by each individual member—the OCA is larger than the sum of its members' weights—in negotiations with other countries or group of countries. A common exchange rate (ER) policy toward outside currencies is required in this case, and this will be easy if a common currency is adopted, but a rather cumbersome coordination effort may be required otherwise, including pooling of members' foreign exchange reserves and supranational management of this pool.

Passing next to the costs, autonomy in the monetary and exchange rate policy of member countries is compromised. The financial integration and perfect capital mobility and effectively perfect asset substitutability render monetary policy impotent. The reduction in policy alternatives can be problematic if wages, productivity, and prices follow different trends in different countries.

Although fiscal policy is potent under flexible exchange rates, the full spectrum of fiscal policy measures may not be available to a member since the majority of countries in the area may adopt targets that run counter to an individual country's goals. The existence of a member with a low inflation and external surplus may force other members with an external deficit and higher inflation to adopt restrictive economic policies which may raise unemployment. Since the mobility of labor is lower than that of capital, it may be the case that regional disequilibria may be exacerbated. Capital will move where it can be most profitably employed (taking into account distance, infrastructure, wages, etc.) and thus may widen the gap between the more and the less developed regions of the area as a whole.

According to Niehans, the currency area debate emphasizes solutions, at opposite ends of the spectrum, to the problem of addressing disturbances, either an inventory (i.e., foreign exchange reserves) approach, while prices (i.e., exchange rates) are kept fixed or an approach of flexible prices (exchange rates) with no emphasis on using inventory (reserves) as a policy instrument. Ideally, an optimal mix of these two pure strategies could be developed which would in general be superior to either of the pure strategies. Seen this way, the choice between flexible and fixed exchange rates then becomes a question to which the theory of the second best can be applied to determine which one of the pure strategies is closer to the optimal policy mix. The compromise systems can be seen as an attempt to find this optimal policy mix although it must be kept in mind that they must tackle the problems of fixed exchange rates that led to the collapse of the Bretton Woods system (destabilizing speculation, export of inflation, etc.) and effectively nullify the criticisms levied against fixed exchange rates by proponents of flexible exchange rates. Furthermore, they must tackle the major problems of flexible exchange rates presented in the beginning of this paper (i.e., the depreciation–inflation cycle, nominal exchange rate movements affecting real exchange rates, high capital mobility undermining the insulation properties of flexible exchange rates, etc.).

Policy coordination seems to be the only unequivocally agreed upon prescription for the aforementioned problems. Such coordination, however, is a technical issue only to the extent that it deals with finding appropriate reward/punishment schedules to deter defection and promote cooperation. Beyond that, it is a political issue involving relative power and perception of threat.

In general, for any region, greater exchange rate flexibility implies an uncertainty in the flow of revenue. Whether flexibility has had a negative effect on international economic activity is dubious. Some analysts, such as Hooper and Kohlhagen, detected no such effect, but their findings have been challenged, e.g., by Kenen and Rodrik (MacDonald 1988). Even though in the long run it is likely that periods with exchange rates favoring one side will alternate with periods not favoring it, so that there will be a tendency for the effect on the flow of profits to even out, one should the gambler's ruin problem which may render the long run, too long indeed.

Finally, it has also been suggested that the adjustment costs in terms of resource reallocation will be small if the exchange rates' variation eliminates possible differences in the cost and price trends and if nominal interest rates simply reflect the different inflation rates. Unfortunately, the former condition has not often held (MacDonald 1988) and the latter, which is the well-known Fischer's condition, suggesting that the real interest rate is constant, has not been corroborated by developments since the early 70s.

The gist of the analysis so far is that the resolution to the apparent trade-offs involves political arguments and political solutions, which, if successful, can accommodate various seemingly contradictory economic arrangements (if internal and external balances can be achieved with both fixed and flexible exchange rates, it is not clear which of the two is the optimal choice). Often, economic, technical debates juxtapose particular arrangements, as if one or the other were the generally appropriate solution, independent of the environment in which they will function. Analysts often fail to see that the performance of an arrangement is limited by, and in fact should be geared to, the political configuration which will harbor it.

Two points of criticism should be made regarding the political economy of much of this analysis. First, although in well-diversified economies' changes in the terms of trade will not have to be frequent because successful exports will offset failures, powerful lobbies may dictate such changes to the government to protect their interests and thus upset the balance pursued. Second, one may ask whether we are necessarily better off if the state bears the exchange risk, whereas in most areas of commerce, the state does not share the risk-facing firms (Gandolfo 1987). The weight attached to individual welfare by the state in reality depends on the individual (i.e., his or her wealth, power, etc.). Domestic commerce receives assistance from the government through its setting price floors and ceilings, setting out the rules of acceptable competition, preventing large swings in factor prices (wages, subsidized energy prices, etc.). Finally, political advantages of enmeshing economically the countries whose rivalries have instigated two world conflagrations in this century may by themselves make the cost worth bearing.

How are we to evaluate the criteria and costs and benefits presented above and especially with respect to the process of European integration? Let us note that criteria perfect factor mobility does not obtain even within one country (e.g., Ohio high versus Massachusetts low unemployment rates) and that language barriers make the situation in Europe even more difficult.

Focusing on McKinnon's degree of openness and Kenen's product diversification criteria, we observe that there is, as we have already noted, an apparent contradiction between them. The ideal candidate for membership to an OCA would be an economy with a high degree of openness (a la McKinnon) and a high degree of product diversification, especially in its exports (a la Kenen). Since such countries are not very common, we may often have to sacrifice one of the two criteria in favor of the other. It may be wiser to attach more weight to the satisfaction of Kenen's criterion since high diversification makes a total drop in exports less likely, and at any rate less subject to capricious fluctuations in tastes, and since even if there is a total drop in exports, there are many candidate products for aggressive promotion and advertising to bring the current account back in equilibrium. Note, however, that in this case, the following expenditure-reducing policies to reduce imports may not be advisable since a highly diversified export sector does not necessarily imply a small non-tradeables sector, and hence, such contradictory policies may hurt too many for the benefit of too few.

It is heartwarming for economists to note that of the six criteria they have put forth, at least four have been espoused and actively pursued by EU politicians. Specifically, all legal barriers to international labor mobility were lifted on January 1, 1988; legal barriers to capital mobility were lifted by the end of 1996 as agreed in June 1988 in Hannover; the convergence of inflation rates has been actively pursued since the 1970s; finally, policy integration—in the economic and political spheres—has been a target for many years, though clearly success in this area has been lackluster, as recent sovereign debt debacles indicate. We will elaborate further on this shortly, suggesting a new criterion related to Kenen's diversification criterion, and in providing a rationale for corresponding technological innovation policies.

4.4 A New Criterion

The crux of the sovereign debt crisis in the euroarea can be seen in this context: Tighter constraints on national fiscal policy have been mentioned as a consequence of monetary integration. Nevertheless, the abolition of trade and capital flow barriers dictates that an increase in the budget deficit would be reflected largely in a deficit in the current account of the particular country with the other members. Monetary integration would preclude resort to exchange rate policies to alleviate the problem, forcing fiscal policy to become contractionary to restore external balance. This forces discipline on a country spending beyond its means, but it fails to discipline surplus countries. Although surpluses and deficits both indicate disequilibria, the former rarely provoke alarm and policies to reverse them. Countries may for instance want to pursue non-expansionary policies to increase their surplus, and such a move would bring trouble to other members and ultimately to the euroarea as a whole. Such a development is quite clearly not merely hypothetical, in view of developments in recent years.

The criteria that have been presented so far fail the test of sufficiency. They do, however, form a set of necessary conditions after amending and merging Mc-Kinnon's "openness" and Kenen's "diversification" as we did above. At this point, we will introduce our contribution to this set of criteria. Our goal is modest; we do not purport to have a sufficient criterion but rather an extension to the set of necessary conditions mentioned above. What is new about this criterion is that it looks not at individual country characteristics but rather at the group of countries forming an OCA.

In an OCA, loss of the nominal instrument (exchange rate and monetary policy) should not endanger stability for any member country—neither internal, in terms of unemployment and inflation, nor external, in terms of balance of payments. Key for such an achievement is the ability of the group to adjust to asymmetric shocks, preserving the stability of its members' economies as well as its own cohesion. The ability to adjust to asymmetric shocks can be the Achilles' heel of the eurozone, in particular, and the pattern of trade among its members obviously influences the response to such shocks. In a few words, this criterion calls for incomplete specialization, increased intra-industry trade among member countries, and technoeconomic convergence toward higher technology plateaus.

As indicated above, technology and research can help buttress a monetary union, through their horizontal, cross-sectoral impact on production structures. This applies particularly to advanced adaptive systems, such as robotic systems that can be used in wide arrays of sectors to enhance performance and competitiveness. This will help European economies to compete internationally by emphasizing their strong card, technology, as opposed to competing on labor cost, on which they could hardly hope to win. Moreover, and more to the point here, it will help mitigate the tensions caused by divergent performances between members of a monetary union, through promoting convergence toward higher technological plateaus and intra-industry trade.

Let us schematically explore the issue (for a more formal model, see the annex): An industry suffers an industry-specific shock, and the country A, which in a regime of complete specialization and inter-industry trade, has specialized in it and is faced with a deterioration in the current account which would require an inflow of capital, generating a surplus in the capital account, to preserve equilibrium in the balance of payments. In an OCA, however, with free capital mobility and perfectly elastic capital flows, and no expectation of ER changes, the usual determinants of short-run capital flows are inactive. In a full monetary union, the interest rate as well as the currency would be under union, and not under national, control. Note that the more realistic assumption of not fully elastic capital flows and hence observable interest rate differentials do not modify the gist of the analysis, since monetary policy would still dictate the cost of money for the country which needs liquidity.

To cushion an economic shock, Brussels can provide transfers or subsidized loans to country A; however, such action could be deemed a "bailing-out" prone to result in moral hazard problems. Furthermore, such transfers, and the expectation thereof, will imply the assumption of a huge fiscal redistributive burden, which would necessitate fiscal union and, effectively the creation of the United States of Europe, to allow Brussels to tax-and-spend through a "federal" budget twenty times its present one.

Alternatively, if the shock reduces employment and wages sufficiently in A and the effects of the shock diminish with time, long-term capital may flow into A in anticipation of high future returns due to low wages and disappearing shock effects. If, however, the shock effects have high inertia and/or result in a permanent shift in the systematic component of demand (or supply), long-term capital flows will not solve the problem. In a "closed" economic area, consumers will have to endure outmoded or reduced quality products until the adjustment takes place (which may take a long time in a protected market). In an "open" economic area, adjustment should be faster and more painful for that industry and its host country. If adjustment is perceived to be too costly, the host country may have to switch to a new specialization, which could signal an even more painful transition for the country and a loss of that industry for the eurozone as a whole (if start-up costs are high, it would be infeasible to keep shutting down and starting up as shocks come and go).

In an intra-industry context, several countries partake in the production of differentiated products of the same industry, products which are imperfect substitutes of one another. If the shock that hits a country is country specific, and not industry specific, we have a situation similar to the inter-industry case examined above. The outlook is somewhat ameliorated; however, when we consider that for the eurozone as a whole, only one variety is lost from the spectrum of differentiated products of the industry in question.

In fact, if labor skills and capital within the industry are not variety specific, countries which produce other varieties of products of the same industry can try to meet the demand for the specific variety produced by the country hit by the shock. Furthermore, if that country is faced with a demand shock against its variety, it can try to carve a new niche with a new variety (remember, we assume capital and labor skills are not variety specific). Finally, losses are smaller since the particular industry represents only a portion of employment and output in the country in question.

If the shock is an industry specific, EU-wide shock (e.g., high energy costs for energy-thirsty products), then it is easier to orchestrate a policy response in terms of monetary or exchange rate policy when many or all members are affected by the shock than if only one country specializing in that industry was affected by the shock.

To recapitulate, the inter-industry case is characterized by higher financial than real diversification, by the increased importance of start-up costs for the preservation of the industry, increased vulnerability of the industry as a whole, increased potential for protectionist pressures in the aftermath of a country-specific shock, high tensions due to wider divergence in targets in orchestrating monetary and exchange rate policy in the union bodies. The intra-industry case on the other hand is characterized by increased real rather than financial diversification, reduced industry vulnerability, increased protectionist pressures in the aftermath of an industry-wide shock, and easier monetary and exchange rate policy orchestration in the face of an industry-wide adverse shock.

This emphasis on intra-industry trade has a corollary—indeed, in the long run a prerequisite—in terms of technology and innovation policies, as one needs to foster wide spectrums of industries in many countries, which can hardly compete globally on the basis of labor costs. The only way to pull it off is to promote technology- and innovation-driven competitive advantages, exploiting the existence or development of varieties that are imperfect substitutes between them. Moreover, the technological drive provides for continuously improving varieties, keeping ahead of the competition, generating wider arrays of varieties, and nurturing consumer interest.

EU policymakers seem to have realized this factor early on, at least in some cases, and its long-term role as a prerequisite. In view of the crises/tensions that followed twenty years later, and with which the eurozone is grappling, those early approaches appear quite laudable. There is evidence of such approaches from as early as the late 1980s and early 1990s, when the Maastricht process, leading to monetary union, was being designed and launched.

This was more evident in regional development policies and in research and technology policies. Brussels would rather trigger internally generated development within each lagging region than shouldering unilateral income-raising transfers to that region, but also because the EU budget is too small—1 % of EU GDP—compared to those of nation-states (in the United States, it is more than 20 % of GDP). To the extent that *specialization may render countries vulnerable to income-reducing random shocks in an uncertain world, and given the union budget's inability to cushion the effects of such shocks with large transfers, specialization without financial diversification is not an attractive option.*

Research and technology development policy was seen, already then, as crucial for economic cohesion (Tecnomics, Background Report on the Impact of the Framework Programme on Economic and Social Cohesion in the European Community, Evaluation Panel on Economic and Social Cohesion, European Commission DG XII, March 1990, p. 8) because: "the less developed regions must have the tools, technology included, to fashion a stronger economic performance in the future. Technological capability is crucial because economic and social cohesion is basically to be obtained by enhancing the less-favored-regions (LFRs) potential for internally generated development. Technology can be the 'engine' for such internal development."

The apparent solution seems to be real diversification on the part of the less wealthy members into industrial activities requiring and nurturing high R&D spending so that technological innovation will guide the home country to development. "If strong economies are sustained by advanced research and technological applications then it has to follow that the economies which are lagging behind cannot develop unless they too are effectively associated with scientific and technological developments" (Policy of Cohesion in the Research and Technology Sectors, European Commission, DG XII, Ref. 9103/88, Nov 17, 1988).

In another report (European Commission DG XII, Gerard Garnier, "Worldwide and Local Technologies," Fast II, Brussels, Oct. 1987, p. 4–5), the issue of specialization is presented in terms of a choice between two different strategies/set of technologies, worldwide and local: "the first strategy is based on the twin notions of an international division of labor and of worldwide technologies. It accepts the postulate that the best strategy to maximize world welfare, and presumably the welfare of every participant country as well, is a strategy of international specialisation. Each country has to specialise according to its comparative advantage, produce in huge quantities for a world market and trade its products against the production of other commercial partners. In practical terms, this means that the country should strive to be the lowest cost producer through large-scale production, but also to become a technological leader in its field." The second strategy, on the other hand, "would stress the national and even the local markets and tie the country's development to the harmonious growth of its regions: this is what the French call a strategy of 'inner redeployment' ('redeploiemement interieur' [sic]) by opposition to the former strategy of outer worldwide deployment." Predictably, the author of that report is not very favorable in his evaluation of the former strategy: "Such a strategy of complete specialization along the lines dictated by the imperatives of an international division of labour has sometimes led to disastrous results. It was the case with several African underdeveloped countries: in the 60s and 70s they restructured entirely their economies which were essentially agricultural economies, based on the production of food crops oriented to the satisfaction of the local needs. These food crops were then abandoned in favor of large-scale industrial crops like cocoa beans, soya beans, peanuts, etc. These crops were entirely exported. The drop in the prices of agricultural products towards the end of the 70s had dramatic results for the foreign exchange earnings and the incomes of these countries. In addition they had to import increasing quantities of food. The end result was their impoverishment and the starvation of thousands of people."

Flexibility is praised as a key virtue (European Commission, DG XII, FAST H. Revermann and P. Sonntag, "Key Technologies," Brussels, Avebury Press 1989, p. 142): "employment flexibility, flexible human 'resources', flexible work organisation, flexible sectoral and geographic mobility: flexibility is likely to become the most spoken principle across the world in the 90s." In addition "success on the market will increasingly be decided by the degree of flexibility possible in production."

Another report (European Commission, "The Regions in the 1990s," COM (90) 609 final, Brussels Jan. 9, 1991) of the time of the launching of the Maastricht process leading to monetary union, suggested (ch. 9, p. 5): "With the completion of the internal market, the exploitation of existing comparative advantages would suggest that the Community's lagging regions deepen their specialisation in labour intensive industries with a low technological content (traditional consumer industries such as textiles, clothing and footwear and other assembly operations), whereas central regions would increasingly specialise in R&D and capital intensive activities. Estimates….suggest that for the Southern regions this could yield substantial benefits whilst minimizing adjustment efforts in the short term." Is this then the report's prescription? The answer is negative: "In the longer run, however, this course of action is not to be recommended."

The report indicates that the loss of the monetary and exchange rate instruments—and the fiscal discipline concomitantly imposed to protect the stability of the monetary union—will leave members, and especially the peripheral ones undergoing structural change, more vulnerable to adverse economic shocks. For this reason, the report (ch. 9, p. 6) recommends that the structural funds be endowed "with a greater capacity to respond more quickly and more flexibly to adverse economic shocks."

More examples come from Commission reports from the early 1990s, e.g., European Commission, PEDIP, Specific Programme for the Development of the Portuguese Industry, COM(90) 205 final, Brussels, May 16, 1990. The report rejects specialization, in favor of real diversification, unequivocally and unabashedly on page 5: "Portugal's industrial development strategy combines investment in fixed and productive capital, in invigorating commercial structures, exploiting precommercial R&D projects and schemes to promote investment in the modernization and *diversification* of the industrial base." Furthermore, "if the country's industrial base is to be strengthened and *diversified*, the excessive reliance on imported technologies must be progressively reduced." In the productivity-driven subprogram, this report indicates (pp. 56–57): "it is the manufacturing of metal products, machinery, and transport equipment which accounts for the largest number of firms receiving assistance (38 % of assisted firms and 41 % of assistance granted); the textile, clothing and leather industries and timber and furniture industries are in second place and it can be said that applications are sectorally diversified."

From the European Investment Bank (EIB) annual reports of this period, in which the Maastricht process leading to monetary union was being launched, it is obvious that specialization was not among the goals for member states. Rather EIB loans appear to help sustain and enhance a diversified production structure in the member states. Spain and Portugal—then the most recent additions to the European Community—follow an interesting pattern in their usage of the EIB lending mechanism, as can be seen from the EIB 1990 annual report. The list of activities funded expands drastically for Spain, and especially Portugal, a few years after acceding to the EU (then European Community).

For Spain, the list of loans that initially is limited to small loans to automobile and telecommunications, by 1990–1991, has added chemicals, pharmaceuticals, electrical appliances, and aircraft manufacturing. In Spain, in the period 1988–1991, the telecommunications industry received 79.4 million ECU in loans, the automobile industry received 567.7 million ECU in loans, the pharmaceuticals industry received 30.7 million ECU in loans, and the aircraft industry received 193.0 million ECU in loans. In the case of Portugal wood fiber, electronics and hotel industries are joined, by 1990–1991, by glass, chemicals, automobile, paper, aluminum, foodstuff, and textiles as recipients of EIB loans. In Portugal, in the period 1988–1991, the electronics industry received 10.2 million ECU, the wood fiber industry 32.7 million ECU, the food industry 11.8 million ECU, the paper industry 195.3 million ECU, and the automobile industry 265.6 million ECU. The pattern of Spain and Portugal is characteristic: Both were successful 1980s' entrants to the European Community, trying to catch up by expanding their industrial spectrum rather than specializing, with European blessing and funding.

We have thus seen how incomplete specialization and intra-industry trade can be justified in a community aspiring to monetary union. Technology and research can help buttress a monetary union, through their impact on trade and specialization, and production structures, placing the emphasis on intra-industry trade. This applies particularly to advanced adaptive systems, such as robotic systems that can be used in wide arrays of sectors to enhance performance and competitiveness. This will help European economies to compete internationally by emphasizing their strong card, technology, as opposed to competing on labor cost, on which they could hardly hope to win. Moreover, and more to the point here, it will help mitigate the tensions caused by divergent performances between members of a monetary union, through promoting convergence toward higher technological plateaus and intra-industry trade.

The preceding analysis can provide a rationale for technology policies favoring horizontally applicable advanced automation robotic systems and intra-industry trade over specialization. It suggests that the EU would rationally process alternative scenarios and evaluate them not only on the basis of general theoretical prescriptions but rather taking into account considerations of discord minimization and linkage of issues during negotiation periods, as well as the new pressures created since the 1980s through the waves of expansions/accessions, and the drive for monetary union—events which have made the EU economy more complex and the cost of failure through discord unacceptably high.

Annex: A Formal Model

We have a set of countries $X = \{x(1), x(2), x(3), \ldots x(n)\}$ and a set of industries $I = \{I(1), I(2), \ldots I(m)\}$. We assume perfect competition among identical firms within each industry, and we represent them by a single firm $F(j)$, where j is an indicator for the industry to whom this firm belongs. The firm's revenue equals payments to labor LB and capital KP and at complete specialization constitutes the income Y of the host country.

$$Y = p\,D = \text{wg}\,\text{LB} + \text{rt}\,\text{KP}\mathrm{d}Y/\mathrm{d}T = p\,\mathrm{d}D/\mathrm{d}T \qquad (4.1)$$

where p is the exogenously given world price for the country's product, wg is wage paid to labor, rt is rent to capital, D is demand for the product, and T is a measure of a demand shock. The income lost from an adverse shock is $A = p \, dD$. Since the interest rate i and the exchange rate are in effect set by the European Central Bank (ECB), the country can borrow the sum A only if expected income $E[Y] = E[p] \, E[D] > A(1 + i)$ over the horizon of the loan. Furthermore, if the expected return in the country's industry $E[r] > i$, then there will be capital inflow. What interest rate should the home country be aiming for during monetary policy deliberations? Assume K flows into the country, of which A is consumed to redress the lost income effect of the adverse shock and $K-A$ is invested in the home country's industry. We must have

$$(K - A)\,(1 + E[r]) > K(1 + i) \Rightarrow i < E[r](1 - A/K) - (A/K) = i' \qquad (4.2)$$

Hence, the home country will strive for an i less than or equal to i' and the tension in the ECB, and the concomitant loss to utility in the particular country and the EU as a whole will increase with the distance between i' and the interest rate $i*$ eventually chosen by the ECB.

In the intra-industry case, many industries are represented in each country and the country's income equals the sum over all j's of $p(j)D(j)$, where j is an industry indicator and $D(j)$ indicates demand for that particular country's variety of products of industry j (as before all revenue is paid to factors of production). For each country, $Y(j)$ is income from industry j and Y is the sum total income for the country from all industries. A negative demand shock T will lead to loss of income $dY(j) = p \, dD(j)$ with the elasticity of demand to that shock assumed to be the same in all countries. The pressure from each country will increase, and the desired i' will decrease the share of that industry's revenue with its country's income increases. As before, countries can borrow the lost income A if the dot product of the vector of expected future prices times the vector of expected future demands exceeds the dot product of the present prices times present demands by at least the amount borrowed plus interest, plus the current deficit in consumption.

Depending on shares of $Y(j)$ in a country's total Y, the i' a country will pursue will vary from country to country. When it comes to deciding on monetary policy, countries will be faced with a trade-off: A low $i*$ will please the countries hit the hardest by the shock and will disappoint those who aimed for a higher $i*$ because the lower i in their cases would bring about inflationary pressures.

In the case of an emphasis on inter-industry trade and concomitant specializations, one country (the one extensively hit by the shock) promotes a low interest rate and all the others prefer a high one. If the high one is adopted, the loss for the country hit by the shock may be devastating and may even lead to withdrawal from the monetary union. If the low one is adopted, then all the other countries will suffer increased inflation.

In the case of intra-industry-trade emphasis, and concomitant specializations, the juxtaposition is not so extreme—some countries' target i' will be very close to the $i*$ chosen and others countries' i' will be more distant, on either side. It seems easier in the intra-industry case to achieve near-satisfaction of more countries without driving any single one to desperation.

In order to examine these claims, we will propose an elementary loss function L which measures loss due to high $i*$ for countries which aimed for a lower i as well as loss due to expected inflation for countries aiming for a higher interest rate than the $i*$ eventually adopted. The loss for the former set of countries increases with the share w of the industry hit by the shock in their income. The loss for the latter set of countries is mitigated by that share w in the former country's income because a high w indicates a high risk for dissolution of the community, seen as a loss by all, including the inflation-fearing countries.

More concretely, one can suggest that bargains are struck which reimburse inflation-fearing countries for their accepting a lower $i*$—the higher the w is, the higher the reimbursement the inflation-fearing partner can extract. We could add terms for other kinds of losses, such as inefficiency due to lack of specialization—it would not change the flavor of the analysis.

The issue at hand involves promotion, or not, of inter-industry-trade specialization within the new context of economic and monetary union. Are there strong new reasons, reasons which are emerging in the context of economic and monetary integration and were inoperative before? We will limit the size of the group of countries as well as the number of industries to two, in order to simplify the analysis. Country 1 aimed for a lower interest rate $i'(l)$ than the $i*$ eventually decided, whereas country 2 aimed for a higher one, $i'(2)$. The constants h and a are positive exponential coefficients of utility loss, and the exponents are overall negative because the bases are between 0 and 1. The exponent a is weighed by $(l - w)$, $0 < w < l$, whose mitigating effect increases with w, for the reasons mentioned above. Each country can specialize in one of the two industries, or opt for producing and trading in both (given the standard assumptions about similar country size, tastes, technology).

Our loss function L is

$$L(w) = (i* - i'(l))\exp(-hw) + (i'(2) - i*)\exp(-a(1 - w)) \qquad (4.3)$$

We will examine how L varies with w. If L is minimized for an extreme value of w (i.e., 0 or 1, or a value <0 or >1), this would favor complete specialization and inter-industry trade. If a value of w between 0 and 1 minimizes L, this would favor incomplete specialization and intra-industry trade. The analysis is symmetric in the sense that country 2 can play tomorrow the role that country 1 is playing today, if an adverse shock hits country 2.

For an extremum, $dL/dw = 0 \Rightarrow$

$$-hy - hw\ln y + az - a(1 - w)\ln z = 0 \text{ where } y = i* - i'(l) \text{ and } z = i'(2) - i*$$

$$\Rightarrow a\ln z / h\ln y = y - hwz - a(l - w) \qquad (4.4)$$

$$\text{set}(a\ln z / h\ln y) = s \qquad (4.5)$$

$\Rightarrow \ln s = -hw\ln y + a(l-w)\ln z$ after taking natural logarithms on both sides

$\Rightarrow \ln s = -w + s(1-w)$ after dividing through by $h\ln y$

$$\Rightarrow w = (s - (\ln s/h\ln y))/(1+s) \qquad (4.6)$$

Let us examine this ratio. We note that $0 < y, z < 1 \Rightarrow \ln y < 0$ and $\ln z < 0$. Furthermore, since both $\ln z$ and $\ln y$ are negative and both h and a are positive, we have $s > 0 \Rightarrow$ the denominator $1 + s > 1$.

If $s < 1$ then $\ln s < 0$ and $\ln s/h\ln y > 0$, and hence, the numerator is less than 1 since we are subtracting a positive number from s, which we have assumed is less than 1. Therefore, if $s < 1$ the numerator $<$ denominator $\Rightarrow w < 1$.

If the numerator is >0, then $0 < w < 1$, and intra-industry trade is favored. Is the numerator greater than 0? Yes, if $a\ln z/h\ln y > \ln s/h\ln y \Longleftrightarrow a\ln z < \ln(a\ln z/h\ln y) \Longleftrightarrow h < a\ln z/(z a\ln y)$. Since $\ln z$ and $\ln y$ would usually not be substantially different and since $0 < z < 1 \Rightarrow za$ must be very small, this condition should often be fulfilled. From $s < 1$, we have that $h > a\ln z/\ln y \Rightarrow$ for $a/za > h > (a\ln z/\ln y) \sim a$ we have the optimal w between 0 and 1, and intra-industry trade is promoted.

The conditions above have the following significance: $s < 1$ means that the loss for #2 is less than the loss for #1 before weighing losses with the exponents which are functions of w. In this case, w is less than 1 because a higher value of w (i.e., $w = 1$) would increase exponentially the already high loss term for #1 and would not offer much in terms of reducing the loss for #2, which is already low. The condition $h < a/za$, $0 < z < 1$, indicates that the larger a is—namely, the larger the exponent of the loss for country 2 is—the wider the range in which h satisfies the inequality.

The above applies when s is assumed less than 1. Note that if $s = 1$, then $w = \frac{1}{2}$—the second-order condition for a minimum is satisfied since $L''(w) > 0$

Now, let us see what happens in case $s > 1$:

if $s > 1$ then $\ln s/h\ln y < 0 \Rightarrow$ in $w = (s - (\ln s/h\ln y))/(1+s)$ the numerator > 0
$$\Rightarrow w > 0$$

For $w < 1 \Leftrightarrow s - \ln s/h\ln y < 1 + s \Leftrightarrow -\ln s/h\ln y < 1 \Leftrightarrow -\ln s/\ln y < h$

From $s > 1 \Rightarrow h < a\ln z/\ln y$, hence $-\ln s/\ln y < h < (a\ln z/\ln y) \sim a$.

The conditions above have the following significance: $s > 1$ means that the loss for #1 is less than the loss for #2, before taking the exponent into account.

w is greater than 0 because a smaller w (i.e., $w = 0$) would increase #2's already high loss exponentially, whereas it would have a smaller beneficial effect in terms of reducing #1's small loss. The condition $-\ln s/\ln y < h < (a\ln z/\ln y) \sim a$ indicates that the range of h which satisfies the condition is reduced as a grows larger—which means that in order for incomplete specialization to be optimal, it should not be the case that $a \gg h$, namely the loss for #2 should not be much greater than the loss for #1.

In other words, when the loss for #1 is greater than the loss for #2, it should not be so overwhelming that any $w > 0$ would lead to exorbitant losses for #1 that #2's benefit from a higher w could never outweigh it (and similarly for loss of #2 > loss of #1 and $w < 1$).

References

Akerlof G, Schiller R (2009) Animal spirits. Princeton University Press, Princeton

Boone P, Johnson S (2010) What happened to the global economy and what can we do about it? The baseline scenario March 2010. (http://baselinescenario.com/2010/03/11). Accessed 15 April 2010

Frankel JA, Rose A (1997) The endogeneity of the optimal currency area. Econ J 108(449):1009–1025

Gandolfo G (1987) International economics. Springer, Heidelberg

Grabbe J (1986) International financial markets. Elsevier, New York

Hallwood P, MacDonald R (1986) International money, theory, evidence and institutions. Blackwell, London

MacDonald R (1988) Floating exchange rates, theories and evidence. Unwin Hyman, London

Mundell R, Swoboda A (eds) (1969) Monetary problems of the international economy. University of Chicago Press, Chicago

OECD memo, May 1986, Grmonetary unione: first examination of reservations to the capital movements code, OECD, Paris

Phillips K (2008) Bad money. Penguin Books, New York

Schloesing L, Stillman E, Bellini J, Pfaff W, Seory J (1973) L' envole de la France dans les Annees 80. Hachette, Paris

Sheets N, Sockin R (2013) Are the Euro-Area Countries an Optimal Monetary and Currency Zone?— A Comparison with U.S. Federal Reserve Districts, Citi Research Economics. (https://ir.citi.com/ tVtjRuFsXuT3pM90FNtdvPoxge5Pn05q2SvG6lnlTDzARnP1rpXXFQ%3D%3D). Last accessed 15 Feb 2013

Tilford S (2006) Will the eurozone crack? Centre for European Reform, Brussels

To Vema, 1 Jan 1989, Athens

Utterback J (1996) Mastering the dynamics of innovation. Harvard Business School Press, Boston

Vergopoulos K, Mouzelis N et al (1986) I Ellada se exelixi. Exantas, Athens

Author Biography

Dimitrios Kyriakou, PhD Former JRC-IPTS chief economist—economic advisor in the Prime Minister's office, Athens, Greece. The views expressed in this article are strictly personal and do not necessarily reflect those of any of the author's employers.

Chapter 5
Perspectives on Technological Developments Applied to Robotics

Clara Pérez Molina, Rosario Gil Ortego and Francisco Mur Pérez

5.1 Introduction

From the very beginning, humanity has pursued the idea of creating an autonomous machine to perform tasks instead of humans in order to make their lives more comfortable. Current robots are autonomous entities capable of receiving information, process it, and make their own decisions to respond on the fly. Nonetheless, there is still a lot work to be done that will require the combined efforts of different disciplines to ensure that robots provide accurate responses to human petitions and desires. In this chapter, we explore the current frontiers of robotics and expose how future robots will benefit new technological developments.

Each section of the chapter is devoted to different aspects related to robots that are closely interconnected and, at the same time, mutually reinforce each other. Section 5.2 is focussed on the robot itself as an entity. The robot is analyzed from a mechanical point of view, and different complementary research areas related to the machine are identified. This is the section more specialized in humanoids; however, not only the physical constitution of the machine is essential, also how the robot interacts, that is the reason why it is call soft robotics, since this represents a more global notion. Section 5.3 emphases on robots designed to move. Those entities are constructed to range over different areas, and we will see how new designs have conquered the world as they are prepared to move throughout the earth, wind, water, fire, snow, or even ride along Mars. Moreover, future

C. P. Molina (✉) · R. G. Ortego · F. M. Pérez
Department of Electrical and Computer Engineering of the National Distance Education University (UNED), Universidad Nacional de Educación a Distancia, Madrid, Spain
e-mail: clarapm@ieec.uned.es

R. G. Ortego
e-mail: rgil@ieec.uned.es

F. M. Pérez
e-mail: fmur@ieec.uned.es

A. López Peláez (ed.), *The Robotics Divide*,
DOI: 10.1007/978-1-4471-5358-0_5, © Springer-Verlag London 2014

designs to promote new missions are highlighted. In Sect. 5.4, more traditional robot applications are considered, mainly manufacturing robots and the challenges they have to overcome will be widely debated. In Sect. 5.5, we explore new trends and technical aspects that robotics will benefit in further designs. Finally, some relevant conclusions with the main achievements and challenges of the current and future robotics are drawn up. We will make ours throughout the text the words of Guizzo and Deyle (2012), "robotics is going through an amazing time, and things should only get more exciting."

5.2 Soft Robotics

As Rolf Pfeifer pointed out (Pfeifer 2010), soft robotics can be considered as the next generation of intelligent machines. The term is used to designate new autonomous machines capable of functioning in the real world, robots with some of the properties that exhibit biological organisms such as adaptability, versatility, and robustness.

Talking about robots, soft can be applied at several levels (Pferifer et al. 2007):

- Soft surface which means soft to touch. This characteristic relies on soft and deformable materials and allows that collisions between human and robots are modeled from a viscoelastic robotic trunk (Lim and Tanie 1999)
- Soft movements which means more natural movements, more humanlike. Soft robots require elastic compliant materials for muscles and tendons or variable compliance actuators
- Soft interaction with people which means smooth and friendly interaction. Soft robots should interact with other agents with movements and behaviors that result natural.

Soft robotics have the proven capacity to convert robots into more adaptable, capable, and safer devices, especially in situations where they closely interact with people in unstructured environments such as homes, offices, and public places (Guizzo 2012).

In the next future, soft robotics will have the potential to lead to a "new industrial revolution" with soft robots being the key to new factory automation technology. Robots with compliant technology will achieve manipulation skills that current robots do not have.

5.2.1 Skin, Hands, and Muscles for Robots

Most of the objects found in our environment are designed for human hands, so one of the grand challenges of robotics is the ability to grasp small and delicate objects with fingertips to adapt to the shape of the object without the need to preprogram the robot. Next generation of artificial hands relies on new developed skin with advance tactile sensors and new method for haptic sensing. In order to achieve that, some groups are building robotic systems that have deformable tissue on the surface, such as Hosoda hand (Takamuku et al. 2008), and anthropomorphic skin-covered hand that enables robust haptic recognition.

Talking about artificial hands, skin is an essential component since it enables the use of object affordance for recognition and control. New adaptive design of anthropomorphic hand structures provides haptic recognition by convergence of object contact conditions into stable representative states, toward this end morphological computation after repetitive graspings are applied (Ikemoto et al. 2011). Visual feedback is not always available, and current tactile sensor output depends on the contact conditions; however, bionic hands with fully covered sensitive skin are increasingly closer to reach the capabilities of human hands.

With the aim of reproducing humanlike behaviors, viscoelastic models are used to model the tendon compliance. In Palli et al. (2012), consider tendons made in polymeric materials where the hysteresis in the transmission system is taken into account as a nonlinear effect because of the plasticity and creep phenomena typical of these materials. The process is characterized by means of a dynamic friction model to consider the effects that cannot be reproduced by employing a static friction model (Palli et al. 2010). These kinds of new control strategies, for the compensation of nonlinear effects and the control of the force that is applied by the tendon to the load, improve the mechanical design of robots and show that the use of new polymeric fibers as tendon materials can be an excellent approach for developing tendon-driven devices (Palli et al. 2009; Jung et al. 2007).

Looking at the full-body robots or humanoid robots, a good example can be ECCE (Pfeifer 2011) which is an anthropomimetic robot developed by the University of Zurich. ECCE stands for embodied cognition in a compliantly engineered robot. It copies not only the shape of a human body but also the inner structures and mechanisms, such as bones, joints, muscles, and tendons using pneumatic mechanisms for developing humanlike actuation systems.

At the University of Tokio's JSK Robotics Laboratory, an advanced musculoskeletal system mimics its human counterpart has been developed. It is called Kojiro humanoid (Mizuuchi et al. 2007). It has about 100 tendon–muscle structures and 60° of freedom. There are also groups that are trying to design novel types of actuators using new materials and mechanisms that can dynamically change their properties.

5.2.2 Legged Machines and Their Control

Since Waseda University developed WL-9DR in 1980 (McMaster 2012), the first robot to exhibit quasi-dynamic walking, many engineers have worked in improving control systems that can adjust gaits and balance the body in real time to allow more versatile walking. This holds true not only regarding the academic environment, but also with regard to trading companies. Sony has developed recently the SDR-4X, a domestic robot capable of handling uneven surfaces and stairs on the fly, and that can sense depth and distance of objects to cope with changing surface heights. Moreover, the Sony humanoid exhibits an incredible flexibility, which allows it even to dance.

The bottom line is how to control that kind of robots, which have so many degrees of freedom, and give them humanlike cognitive features. The way the body and muscles interact is difficult to model using classical control methods; since there are lots of nonlinear behaviors, the tendons have static and dynamic friction and the mechanics are not precise. The ability to perform a steady unhurried movement requires the actuation of multiple muscles that need to be coordinated, actuated to varying degrees. At this point, learning becomes essential; robots need to figure out things by themselves. This is where the concept of embodiment (Pferifer 2007) appears.

Another approach in the generation of motion is the design of algorithms that respond to tactile stimuli of the human being as the strength thereof. It tends to maximize the contact points of the robot with human beings. In Schmidt et al. (2006), it performed a robotic arm that will follow the movement of the human being as he moves. The drawback of this robot is that at the end of the operation, the robot does not store any results, i.e., cannot use any data as feedback for future behavior.

As it has been mentioned previously, control systems that take care of everything the robot does can work well in structured environments, such as factories. This can be considered as the classical approach. However, when we have environments that are constantly changing and unpredictable interactions, we need to go beyond using new control systems. In fact, we need a new notion of control.

For instance, talking about a robot that walks on two legs, we have to forget the classical zero-moment point method, planning how to actuate each joint. Researchers are considering new points of view such as passive dynamics where part of the control is outsourced to the physical dynamics of the robot. There is not a clear separation between control and the controlled. That is the reason why some experts use the term of orchestration rather than control (Haapasalo and Samuels 2011).

Under embodiment approach, every action has a consequence in terms of patterns of sensory stimulation. The plan is to let the robot, which is equipped with many sensors, explore, and learn on its own. In this way, sensors provide useful information on how effective its motor signals were for a particular movement. Thus, from the feedback it gets, the robot can figure out its own dynamics. We need biomechanical systems with similar properties to our own bodies that are compliant, reactive, and have control distributed throughout its subsystems.

The goal is to search for global parameters rather than trying to control every detail of the movement; otherwise, controlling a system with potentially infinite degrees of freedom becomes an unaffordable work. Those parameters should be allowed for tolerating changes on the fly, leaving the details to the morphological and material properties. This kind of soft technologies is also a part of soft robotics.

The long-term objective for today robotic engineers is to develop a robot able to help people in their daily lives, which means that the robot has to be able to operate in our environment. That is the reason why new humanoid robots should to be designed to walk and move just like us. Recent designs and materials with novel characteristics will provide an optimal elasticity of the muscle–tendon system that involved in an adequate shape for building legs and arms, which will result in best-performing robots with enhanced walking way or grasping a hard object way, for taking an example.

Soft robots provide the opening up of new horizons to robotics. The appearance of active soft materials, that embed electrical or mechanical functionality in materials that are inherently soft, will allow researchers face new challenges in the very next future.

5.2.3 Robots as Partners and Accessories

Humans and robots can collaborate to perform practical tasks. A definitive step toward robots migrating out of factories and academic laboratories into our everyday lives has been taken.

According to the US National Robotics Initiative (NRI), corobots is the term used to describe robots that must be safe, relatively cheap, easy to use, available everywhere, and able to interact with humans to leverage their relative strengths in the planning and performance of a task. Those kinds of robots will behave as partners, accessories, coworkers, or colleagues among other things. However, there is a considerable way to go, and major efforts are still necessary in human–robot interaction to develop a strategic approach that results in increased levels of activity in human–robot interaction in the coming years.

Implementing new robots in human environments, especially the public, carries high spatial resolution sensors, and complete coverage of the robot's body also must consider the interpretation of human intentions. In the end, the goal of these interactions will facilitate communication between human and robot. There are some studies that try to infer human intentions through touch gestures on the robot. Through a preprogrammed mapping, different actions can be classified into families' defined paths, and this classification influences the selection of which path to be followed. The key point lies in the selection of behavior within industrial environments where the robot can operate at different modes without reprogramming (Ikemoto et al. 2011; Pfeifer 2011). In Mizuuchi et al. (2007), it develops an extensive tactile alphabet which has multiple contacts with multiple fingers to use teleoperation directly or to control the robot through the selection of behavioral commands.

Another important technical barrier is the lack of appropriate safety procedures implemented in robots. When robots interact with humans, compliant actuation is a critical issue. Making robots that have soft touch is key to the future, where humans and robots will share spaces and collaborate closely. Coming soon, we will see numerous improvements in compliant actuation and tactile sensing technologies, as better series elastic actuators and tactile skin. In fact, that kind of new compliant systems has nothing to do with technologies based on electrome-chanical motors.

The ability to perceive from different modalities, along with improved spatial resolution, provides more complete data and at the same time allows the development of more sophisticated behavior by robots. In this regard, it is known that large amount of information is contained within the physical contact. The detection of the force determines the presence of a contact; the temperature data and electric field determine if the contact comes from a human being; the vibration provides an indication of the dynamic of the contact; and pressure is an indication of the magnitude and duration of the contact. The idea of combining multiple sensors is closely related to the issue of capturing the multiple perceptions that make human skin. Human skin is able to detect temperature, touch, pressure, vibration, and damage. Beyond the combination of multiple contact sensors, another open area is local data tactile combined with the information from sensors which detect remotely. The soft touch sensors are similar in sensation as well as temperature as humans. The android Repliee Q2 is able to distinguish a wide range of contacts, from a simple touch to a hit, which is based on motion capture of human beings to emulate later (Ishiguro 2007). On the other hand, Geminoid HI-1 aims to duplicate a particular human being through tactile feedback and generate natural interactions with other humans (Nishio et al. 2007).

Present systems (Nuria and Oliver 2000) integrate together a real-time computer vision and machine learning system for modeling and recognizing human behaviors in a visual surveillance task (Oliver et al. 1999). The system is particularly concerned with detecting when interactions between people occur and classifying the type of interaction. These systems combine top-down with bottom-up information in a closed feedback loop, with both components employing a statistical Bayesian approach (Oliver 2000). The coupled hidden Markov models (CHMMs) are shown to work much more efficiently and accurately than hidden Markov models (HMMs). Finally, to deal with the problem of limited training data in these systems, a training system is used to develop flexible prior models for recognizing human interactions.

Other research being undertaken is the study of human actions upon contact with the robot and the encoding of an adequate response by the robot. This research gives rise to a new term: Physical InterFerence and intended contACT (PIFACT), (Iwata et al. 2000; Iwata and Sugano 2005). Efforts focus on making adaptive PIFACT where the robot analyzes human interaction and the performance of its own work, which includes any active or passive interaction, expected or unexpected. PIFACT interactions are reflected in the humanoid 52-DoF WENDY, where interactions are classified according to the rights of robots and humans. This humanoid has the ability to maintain its position when it is disturbed.

Shortly, compliant and soft-bodied robots will emerge. Just to take a current example, we can consider the iRobot's Hexapod JamBot (Scott 2012), created by means of a technology developed at the University of Chicago, which is based on particle jamming actuators. This robot platform is composed of six jamming modulated unimorph segments for legs, while the body has six individually addressable jammable chambers. This structure allows the robot to be completely soft in an unjammed state or rigid in a jammed state, or the Ant-Roach Pneubot (Coxworth 2011) designed by OtherLab, a six-legged walking inflatable robot whose weight is slightly under 70 pounds and is made of fabric and pneumatic actuators.

5.2.4 Entertainment and Toy Robots

There are diverse offers of robots which are oriented to play, and the price becomes not to be a problem. National Institute of Advanced Industrial Science and Technology (AIST) from Japan introduces Life Innovation with Therapeutic Robot, Paro. We describe in later health-care robots.

AISOY Robotics from Spain aims at radically modifying the stage within the educational robot market, by providing a new generation of low-cost programmable social robots with emotional capacities that may perceive sense, evolve emotionally, make appropriate decisions, and have a natural, humanlike interaction with users to fulfill its mission.

Another example is Aldebaran Robotics which will offer humanoid robots to the general public in the near future. They created the NAO robot which has already become an internationally recognized robotics platform used in education and research. The ultimate human dream of creating an artificial companion to assist humans is no longer only science fiction but a realistic answer to the needs of an aging society. Aldebaran Robotics develops applications to make this dream reality by conducting research in areas such as autistic child therapy, human–robot interaction, and personal robotics. Futuroscope Company combines images with physical thrills and excitement which makes possible an immersive experience, interactive visits, open-air games. Its most famous attraction is dance with robots. They are robot arms that move people in six directions and can accelerate up to 3G.

A huge area in toy robots is that orientates to programmable personal robots for educational institutions at all levels. Thus, their robots can be used in teaching program. Karots by Violet Cop is funny, educated companion. He can speak, see, listen, obey, and wiggle his ears. He looks like a kind rabbit. Moreover, his entertainment, i.e., he can make free calls, makes the Internet tell us about all the things we want to know (weather reports, news, sports, arts, stock prices, horoscope, TV guide, and all the RSS feeds that we want), follow friends' Facebook status and Tweets, keep an eye on our home and family (being alerted whenever someone enters our house), and listen to our favorite music. Even, we can learn a language.

More than 25 years of experience assures LEGO to cover diverse important curriculum area. His philosophy is based on a hands-on learning approach that actively involves students in their own learning process. For example, with LEGO

MINDSTORMS students learn to design, program, and control fully functional models. They use software to plan, test, and modify sequences of instructions for a variety of life-like robotic behaviors. And they learn to collect and analyze data from sensors, using data logging functionalities embedded in the software.

Marorobot develops a variety of robot called Medabots, 3D character building toys, whose purpose is to learn and enjoy building own robot mechanism. So Marorobot is a company oriented to educational material with a low price. Follows this idea, minirobot developed Metal-Fighter, a humanoid robot for educators, students, and robotic hobbyists. This robot can walk and run and do flips, cartwheels, and dance moves.

PaPeRo by NEC is intended to communicate and interact with people and IT devices, recognizing familiar faces, talking and reacting to compliments. And Reeti by Robopec is a communicating robot with a fully animated head, equipped with cameras that allow it to perceive and express emotions. Reeti can be used as a media center PC. Another example is RoboRobo Corp. that has developed elementary robot education courses and manufactured educational robot kits, toy robots, and competition robots.

There are a huge number of robots for games or educational purpose, and many of them are very cheap, such as MyKeepon, a chicken pet robot, is a nice robot that can recognize music or claps and can dance with the perfect beat. Pleo Reborn is another pet robot and is a dinosaur who born and grown up in a unique way in its species. MyButterfly is as its name butterfly that moves like it and flies with a few simple taps on the lid. Or HexBug is a robot that moves and looks like a real bug. As we can realize, there is a huge amount of different robots for entertainment.

Other applications are being developed that are an intersection of tactile feedback on those that affect the robot and on those that report the state of the human being. The robot ball room dance partner robot (PBDR) (Aucouturier et al. 2008), acts as a fellow human task, particularly in the realization of ballroom dancing. The man guides the movements of the robot, which performs the contact to infer human intentions through the force generated in the x and y axes together with a force point dataset, a coding human motion model based on force and a series of data obtained with a long range laser (Takeda et al. 2007) are data used to predict the next dance step. Such robots as well as their entertainment may be able to know when a human being can fall down (Hirata et al. 2008).

On the other hand, sex robots will be the future of buying sex and of course robots as many researchers have already predicted. In the near future, around 2050, we will be able to go to nightclub and be served by virtual creatures that will be different in ethnicities, body shapes, ages, languages, and sexual features. This innovation will have an impact on the actual human prostitutes, since many sexually transmitted infection and human sex trafficking will be avoided.

5.3 Mobile Robots

5.3.1 Self-Driving Vehicles

Autonomous vehicles have proliferated in the last few years. For instance, Google has extensively demonstrated its self-driving car (Toyota Prius) (Markoff 2010). The countdown for driverless technological vehicles populating the streets has started recently, in 2012, when Nevada became the first US state to permit autonomous cars to be legally driven on public roads.

We can say that cars are becoming more robotic. Nowadays, autonomous driven features are already included in existing regular mass produced cars. In fact, some models allow the driver sets both the desired speed and the safety distance to the vehicle ahead and even exhibit driving assist function that keeps the car centered on its lane or another function that can park the car all by itself.

Two factors must be taken into account in the future and expansion of these vehicles: How far are these vehicles capable of detecting elements and if they are fast enough to react to short distances. Some experts pointed that in order to help autonomous cars and robots navigate, we should start to map in detail our environment.

5.3.2 Unmanned Autonomous Vehicles

Unmanned vehicle technology has advanced to the point where platforms perform persistent surveillance missions far from remote operators (Squyres et al. 2004). Combining unmanned platforms with real-time environmental models enables the adaptive sampling by autonomous robotic sensor systems of data essential for examining the fundamental behavior of complex phenomena. Robotic sensor systems have been deployed successfully in a variety of applications that include hurricane sampling (Lin and Lee 2008), underwater ocean observation (Leonard et al. 2011), pollution monitoring (Ramana et al. 2007), tornadic storm penetration (Elston et al. 2011), water condition mapping (Weekly et al. 2011), and harmful bloom tracking (Smith et al. 2010b).

While traditional sensor networks only provide fixed monitoring points without the means to adapt to changes in the surrounding environment, a new generation of sensor networks has emerged as a new tool to collect spatially dense information in real time from natural environments. They are the solution to meet data requirements in collecting data from diverse events. At a global scale, remote sensing satellites are typically used, while at the regional scale, fixed monitoring stations are mainly employed. Sensor networks solve limits in data collection during significant events such as hurricanes and floods.

Unmanned autonomous vehicles have to resolve many problems (energy efficiency, harvesting capabilities, and endurance of robots among others) to allow robots to operate more accurately and reliably in a range of natural and manmade environments. An example is the energy storage, particularly for autonomous underwater vehicles (AUV) and unmanned aerial vehicle (UAV) by onboard energy storage or the amount of solar panels for autonomous ground vehicles (AGV) and autonomous surface vehicles (ASV), since we must take into account that there is an inherent limit to the amount of solar panels and batteries that can be carried for energy capture and storage.

The idea of these robots in future is according to (Bongard 2011) copy physiological and neurological systems observed in animals and build them into robots. Those innovative systems will incorporate computer programs that copy the dynamics of biological evolution and replay them in a virtual space with numerous generations of synthetic creatures. The resulting algorithm yields ideas for robots that have optimized their neurological structures, their behaviors, and body plans, over many generations of being tested by virtual evolution, instead of human guesswork.

We can distinguish the UAVs for marine systems and terrestrial. The range of application of unmanned autonomous vehicles is diverse, since they are prepared to have different uses. On the military side, there are different applications such as perform reconnaissance and survival, object detection, real-time view of the battlefield, air combat, sensors for chemical and bacteriological warfare, detection and neutralization of minas, camouflage smoke dispersion, fight against terrorism. Similarly, UAVs are used in civilian applications such as fire detection, commercial fishing, identification and recovery of underwater objects, rescue, border control, inspection and preventive maintenance of offshore facilities, surveillance in urban areas, monitoring of highways, inspection of underground structures of small diameter, inspection in nuclear areas, weather, among others.

Beyond these tasks, the future of robots aim to hostile environments or any other environments such as cleaning up a toxic dump.

5.3.2.1 Examples and Trends in Environmental Robotics

Remote environmental characterization in hazardous circumstances represents one of the most acute application areas where research efficiency is rapidly increasing. Thus, some researchers argue that we are in the era of environmental robotics (Dunbabin and Marques 2012).

There is a diverse range of environmental monitoring applications within the marine, terrestrial, and aerial domains. An advantage of using robotics in environmental sciences is that they allow the monitoring and sampling of events that are too dangerous, or impossible, for humans to undertake. A key attribute of environmental monitoring is the measurement of relevant environmental variables, being physical, chemical, or biological. Trends focus on a multidomain (air, land, and water) environmental robotic systems.

There are many practical designs in aerial environmental robotics. Small unmanned aerial vehicles (UAVs), such as quadrotors, increased significantly in the past few years. Currently, we can find Air Swimmers remote-controlled blimps, small smartphone-controlled helicopters, and Parrot's AR Drone quadrotors, among other flying devices. Not only academic researchers but also enthusiasts and curious are involved in this kind of environmental monitoring upon becoming inexpensive robot platforms.

Microunmanned aerial vehicles (MUAVs), especially quadrocopters, are suitable for gas distribution mapping as they can be precisely controlled and equipped with a variety of sensors. Furthermore, quadrocopters have the ability to hover over a certain position, which allows more informative gas concentration measurements with slowly responding and recovering chemical sensors in comparison with a plane. Environmental monitoring using gas-sensitive unmanned aerial vehicles addressed the problems of measuring the spatial distribution of chemical plumes and searching for the emission source.

Another kind of flying platforms is the unmanned aircraft systems (UAS), and they are basically used for sampling storms and related phenomena. More than 50 years of investment and advancements in remote weather-sensing systems (satellite-based as well as ground-based radar) have resulted in remarkable capabilities; however, these systems cannot deliver observations to meet current requirements for timeliness, positional precision, and the acquisition of data that can only be obtained in situ. Highly mobile observation systems are needed to deliver in situ data that are critical for the verification and validation of current models and simulations, so currently, UASs are essential for sampling severe local storms. In short, UASs are potentially useful for the study of a wide variety of atmospheric phenomena and processes, including thunderstorm outflows and gust fronts, landfalling hurricane boundary-layer circulations, planetary boundary-layer fluxes (particularly those relevant to climate dynamics), atmospheric responses to fires, pollutant dispersion, and terrain-driven circulation systems.

Furthermore, the ability of robotic platforms to perform large-scale space missions is well known. In fact, ongoing projects are now bearing fruits. In the Fig. 5.1, we can see a high-resolution image of Mars, thanks to the unmanned vehicle called Opportunity from NASA, which uses the solar energy.

The aerospace industry stays abreast of new scientific discoveries and ever-improving technology to enable high-level space missions with enhanced safety and economy. The space activities of humankind in orbit, on the moon, and on other planets are expected to be increased in years to come, and new autonomy robots will play a fundamental role.

On the other hand, robotic and autonomous systems are playing a crucial role in improving our understanding of the world's oceans. The difficulty in observing the oceans in detail using remote sensing has led oceanographers to employ an increasing number and variety of in situ autonomous sensing systems. There are many examples of autonomous underwater vehicles (AUVs) for oceanographic measurements, such as the REMUS (Allen et al. 1997) and AutoSub (Griffiths et al. 1998). These AUVs have global positioning system, GPS-denied navigation,

Fig. 5.1 Late afternoon
shadows at Endeavour Crater
on the *red* planet, a crater that
spans 22 km in diameter.
Courtesy NASA/JPL-Caltech

mission and path planning, which allow capturing of accurately localized environment data in harsh environments over extended periods of time.

In terms of increasing the endurance of robotic boats, Rynne and von
Ellenrieder (2009) developed a sailing autonomous surface vehicle (ASV) for
sustained ocean operations, with a hybrid solar and sailing vehicle with applications such as tracking marine mammals across vast oceans (Klinck et al. 2009).
Argo float program (Gould et al. 2004) is incredibly energy efficient, with design
endurance exceeding 1,500 days as they drift in the ocean currents. Modern float
array takes measurements of temperature, salinity, and depth. More than 3,000
Argo floats are currently adrift in the world's oceans, and their observations are
used to validate and drive global-scale ocean circulation models. Thus, ASVs are
used for the monitoring of water resources, such as controlling threat of mass
proliferation (bloom) of noxious cyanobacteria. The ASV is designed to combine
the ability to take measurements within a range of depths and accurate localization
provided by the global positioning system (GPS).

ASV as well as AUV has a large range of applications as they can access areas
that are potentially dangerous for humans. The four most important uses of these
platforms are military application (Elkins et al. 2008), structure inspection, shipwreck surveys (Bingham et al. 2010), and ecological studies (Smith et al. 2010a;
Dunbabin et al. 2004).

In addition, it should be noted that underwater communication is difficult and
allows only for limited transmission rates. Both localization and communication
can be achieved easily on surface waters, and therefore, more sophisticated

applications have been developed. Military and defense applications deploy ASVs to patrol shorelines or harbors. Elkins et al. (Elkins et al. 2008) have assembled an ASV system that operates on relatively large motor boats and possesses a sophisticated set of sensors to detect obstacles or target boats, which can then be followed. Only the ASV constructed by Dunbabin et al. (Dunbabin et al. 2009) is able to take measurements at depths down to 5 m while moving. The aquatic networked infomechanical system (NIMS-AQ), developed by Stealey et al. (2008), is capable to measure at deeper positions; however, its motion is restricted to straight lines. To assess the variations on both vertical and horizontal axes in a relatively short time span, (Hitz et al. 2012) designed an ASV to carry a custom-made winch so that it can take measurements as deep as 130 m while sailing to realize a cross-sectional measurements in the lake.

As we have stated before, most of the autonomous vehicles are used in hostile environment. One of them which are booming is volcanic environments. Most of the measurements necessary for a comprehensive analysis of what is taking place inside a volcano should be taken in the proximity of the craters. The main goal of using robotic systems for volcanoes is to reduce the level of risk involved for volcanologists who are working too closely to volcanic vents during eruptive phenomena. In order to predict dangerous eruptions, they need broadly the following information: visual measurement of craters and domes; thermal image to avoid destroying the own system of the robot; gas analysis and sampling; analysis the terrain which is very heterogeneous and temperatures.

Robotics in volcanology is not widespread because of the technical difficulties involved in developing robots suitable for harsh environments and because of the lack of substantial economic resources as those that are available to the military or space sectors (Guccione et al. 2000). In any case, several examples can be given as Dante II, a robot that is able to perform measurements in live volcanoes. The robot was an eight-legged frame walker, with the pantographic legs arranged in two groups of four.

On the other side, most volcanoes can be found underwater and the interest of exploration underwater volcanoes has increased and several research groups have adopted underwater robots to survey the depths of oceans, such is the case of the autonomous benthic explorer (ABE), a robotic underwater vehicle used for exploring the ocean to depths of 4,500 m. Another example can be the AUV Sirius, a platform that includes detailed, high-resolution benthic imaging, multibeam swath bathymetry, conductivity, temperature, depth profiles and fluorometer data measuring chlorophyll-a, colored dissolved organic matter (CDOM), and turbidity at the benthic reference sites. These AUV systems are designed to help characterize changes in benthic assemblage composition and cover derived from precisely registered maps collected at regular intervals. This information will provide researchers with the baseline ecological data necessary to make quantitative inferences about the long-term effects of climate changes and human activities on the benthos.

5.3.3 Telepresence Robots

Telepresence robots are mobile machines that act as your stand-in at a remote location. We can expect to see more telepresence robots around offices and homes in the coming years.

The Web has enabled a growing number of tools for us to work remotely and even collaborate with one another from distant locations. While online meeting services like GoToMeeting, Fuze, and WebEx are effective at letting us communicate with one another, they do not quite compare to the experience of actually being there. The idea behind telepresence robots is go to the office without physically leaving home.

The problem is that commercial telepresence robots are pricey. Even with the current price tag, they may well be worth it for many companies, for whom telepresence robots may increase productivity and help keep other costs down. The following are merely some interesting examples of commercial devices: the QB telepresence robot by Anybots, Vgo from Vgo Communications, and Jazz from Gostai. On the other hand, there are initiatives of open source mobile telepresence such as Sparky Jr. (Cornblatt and Sparky 2012), a telepresence robot built on an iRobot Create base that let us build our own telepresence robot.

An example of the approach of telepresence in different environments in addition to office is the telepresence robots' home with post-op patients that was conducted by Children's Hospital, Boston, via a pilot program that integrates telepresence robots into regular post-op care regimen. Using five robots made by Vgo Communications Inc., doctors and nurses are opening a direct line of communication and observation between themselves and patients even as they recover at home.

5.3.4 Health-Care Robots: Older People Assistance

Bearing in mind that estimates point out that more than a fourth of the total population in the developed countries will be over 65 in the next 20 years (in fact, today this is almost a reality Japan), the increasing demand for healthcare workers will explode. Robotics has become a promising technology to meet this challenge. A new generation of care robots or carebots is emerging to help the elderly and chronically ill to remain independent, reducing the need for human carers and the demand for care homes (Jervis 2005). Carebots could help in several ways such as addressing cognitive decline (reminding patients to drink or take medicine), collecting data and monitoring patients, avoiding emergencies as heart failure or high blood sugar levels, assisting people with domestic tasks, among others.

A carebot is a special kind of robot with specific features. It must be easy to control, to allow the caregiver, a person without a technical profiling, to concentrate on the patient. In addition, the way of human–robot interactions are carried out must be significantly different because the human has a more passive behavior (Ohmura and Kuniyoshi 2007).

We can find many examples of high-tech carebots that are currently being performed successfully, and taking into account that developed countries have put a lot of resources in this area, predictions indicate rapid growth for this kind of service and its quality.

For instance, RI-MAN (Mukai et al. 2008; Odashima et al. 2006, 2007) is a humanoid that is capable of moving humans in rescue efforts, mobilizing elderly people, or monitoring tasks in the hospital. The RP6 (Jervis 2005) is able to mimic the head movements of humans in conversation, helping make face-to-monitor contact with the patient more intuitive, and this two-way audiovisual has been demonstrated to be critical to patient acceptance. The RP6 is a remote presence robot that operates over a secure broadband Internet connection using wireless protocol which enables 25 frames per second video display (equivalent to VHS). Through these robots, doctors can visit patient on a ward, making the process lee disruptive for them; moreover, they may support earlier patient discharge and better use of doctors' and patients' time. This new generation of robots has demonstrated great advantages over fixed telemedicine systems in terms of costs and benefits.

HAL 5 is an exoskeleton developed at Japan's University of Tsukuba that can give patients more mobility and help their caregivers to lift and move them safely. ASIMO is another example of bipedal robot developed by Honda, and this robot allows patients to execute a task without having to move their selves. ASIMO is an aid to the nurses, taking care of the heavy aspects of their tasks and, at the same time, allowing them to spend more time with the patient.

There is also a more specialized field of action robots that focus on psychological research. This is the case of Paro robot (Shibata and Tanie 2001), which looks like a polar seal. According to the contact that is applied, the robot is able to develop a finite number of pulses. Interactions depend on the readings of the sensors, human contact is classified, and adaptively, the robot's behavior can be modified. A trend in the use of robots for therapy is developed in (Stiehl et al. 2005), where thorough human contact is studied and affective functions causing a communication or emulation of an emotion are analyzed. These approaches could be useful in cases of certain diseases, such as anxiety or loneliness.

Regarding the interaction approach to contribute to the robot's behavior to infer the intentions of human beings, most notable is the need for development or improvement when the robot interacts with people who are not experts in robotics. Both the Paro and Huggable robots, which inferred human intentions through contact that performed on them, are robots with a similar design for a pet. Advances in therapeutic activities with real animals is a very promising area in robotics to simulate these real animals in groups of people with psychological problems.

5.3.5 Biorobotics

As Stephen Spielberg said: "There is no such thing as science fiction, only science eventuality." We are nearing the time when technology envisioned in some science fiction becomes reality, although there is still a long way to go there, the trend is upwards.

Biorobotics lies between computational theory and mechatronic engineering implementation, where also recent rigorous mathematical results contribute to enable engineering advancements. Not only human features try to be emulated also animals ones. For taking an example, we can use the Snake Robot (Degani et al. 2010) developed by professor Choset at the Carnegie Mellon University. It is a highly articulated system which can exploit its many internal degrees of freedom to thread through tightly packed volumes, accessing locations that people and conventional machinery otherwise cannot. In terms of control, the robot must plan in a multidimensional space, one for each degree of freedom. Choset's approach uses a retract-like structure of the space, which reduces planning to a one-dimensional search.

Areas such as robotic prostheses and brain machine interfaces are real promises. Exoskeletons have been taken out of the laboratories. Berkeley Bionics has ready to sell its robotic suit, Ekso Bionics, to rehab clinics in United States and Europe, hoping to have a model for at-home physical therapy very soon. These exoskeletons aim to treat spinal cord patients. A person just has to balance his upper body, shifting his weight as he plants a walking stick on the right. A physical therapist will then use a remote control to signal the left leg to step forward. In a later model, the walking sticks could have motion sensors that communicate with the legs, allowing the user to take complete control.

Researchers from Johns Hopkins University and the University of Pittsburgh have testing a brain implant that allows patients to control an advance robotic arm by means of just their thoughts.

Other example is bionic vision, a brain implant or cortical implant provides visual input from a camera directly to the brain via electrodes in contact with the visual cortex at the backside of the head. A computer is used to process the sensory streams, as is typical for a brain–computer interface (BCI).

On the other hand, robotics is also applied to surgery to perform procedures inside a patient's body. Intracardiac surgery is one of the most emerged fields. In Harvard Biorobotics Lab, a robotic catheter (Kesner and Howe 2011b) has been developed whose goal is to create a platform technology to enable minimally invasive surgery on the inside of the beating heart. The system uses a real-time 3D ultrasound imaging that virtually stabilizes the heart motion, allowing clinician to perform surgical repairs as if the heart was stopped.

5.4 Manipulation Robots

5.4.1 Manipulation Robot Arms

Robot manipulation is a crucial research field in robotics. Manipulation of robot arms' performance needs help and support of basic fundamentals of robotics: kinematics and dynamics, motion planning and control, and higher mathematics and artificial intelligence. Most of the modern theories in robotics arise from the classical concepts of robot manipulation. In addition, this field constitutes the more direct application to industry at the moment, since robot manipulators are still a growing market where technology is now being applied beyond conventional areas such as small industries. In fact, most of the new research on interactive robots is centered around it. In this sense, the US Defense Advanced Research Projects Agency (DARPA) launched at 2010 an Autonomous Robotic Manipulation program (Guizzo 2010) which aims to achieve both software and hardware developing that would enable robots to autonomously perform complex tasks with humans providing only high-level direction.

Manipulators are, in terms of robot design, basically kinematic chains with links and joints that have to be controlled. The robot arm control is based on the theory of kinematic; in fact, it is used in both position and velocity domain. While classical robotic arms exhibit structures to make it easy to avoid complex calculations, modern day robots are equipped with powerful computers that do not provide any restrictions on the structure design (Kemp et al. 2007). The new generation of manipulators has been produced with new materials and algorithm techniques for their control.

It is reasonable to expect that the overall cost of a robot falls due to the popularization of technology improvements. Nowadays, this assertion is a half-truth, since it is true for some aspect, such 3D sensors, but not in general, as actuators show. There are several programs focused on new hand designs with high degree of freedom, which are the bases for low-cost compliant autonomous manipulation systems.

5.4.2 Factory Robot Helpers in Manufacturing

Factories represent the typical controlled environments. There, the world can be adapted to match the capabilities of the robot. Within a traditional factory setting, engineers can ensure that a robot knows the relevant state of the world with near certainty, as the robot typically needs to perform a few tasks using a few know objects. However, environments are often too simplified in order to focus on other areas of interest. New autonomous robots have demonstrated to perform complicated manipulation tasks relying on new techniques.

Of course, a factory can be also considered as a human environment, where working robots have to be aware of their surroundings and must meet real-time constraints to interact with people avoiding collisions with humans and other obstacles such as other autonomous robots. The design mode of the robot must be able to recover from a disturbance without endangering any human being and at the same time not detrimental to the robot. They are also interactions should always take into account as they operate in environments where there are both, humans and robots, and the interaction between them is inevitable. For example, the SDR-4X II humanoid (Kuroki et al. 2003) performs a capture detection to avoid pinching or hurting a person. It also detects when someone rises through a series of touch sensors on the handle, and it can be safely transported by humans.

Nowadays, outside of controlled settings, robots can only successfully perform sophisticated manipulation tasks when operated by a human operator. Researchers are pursuing a variety of approaches to overcome the current limitations of autonomous robot manipulation in factories. In fact, any kind of real environment has a big number of challenging characteristics that usually go beyond the control of the robot's creator. Mainly active perceptions including vision, enhanced autonomous learning techniques, platform design and control are the most challenges facing the field.

One option that we find these days is software packages that can be integrated with the robot operating system (ROS) that allows robots to build up a representation of their environment using data fused from three-dimensional and other sensors, generate motion plans that effectively and safely move the robot around in the environment, and execute the motion plans while constantly monitoring the environment for changes.

It is clear that there is a huge need for flexible, capable, and safe manufacturing robots, a new generation of industrial machines different from the big and expensive current manipulators. The Foxconn Company announced in 2011 that it was planning to add 1 million robots (double the current industrial robot population) to its assembly lines over the next three years.

The next-generation factory robots could be just a pair of arms or hands, or other kinds of manipulators, which do not even have to look humanlike. It probably makes no sense to put humanoid robots in factories. However, those manipulators will exhibit advanced designs using new materials and their movements will be controlled by new algorithms.

5.5 New Features and Trends

In this section, we will see some of the emerging research trends for facing new challenges in robotics.

5.5.1 Microrobots

The implementation of microrobots needs the conjunction of expertise in microfabrication and microsystem design. In recent years, biologically inspired systems with high-performance aerial, ambulatory, and aquatic properties have been developed.

As the characteristic size of a robot decreases, challenges for successful questions in general well answered for large robots appear. The hardware to be used for mechanisms, sensor, and computation has to be custom-manufactured in order to match the individual robot's need exactly. All the process requires new ways to design and prototype.

Microrobots constitute a revolutionary tool that can be used for search and rescue operations, assist agriculture, environmental monitoring, and exploration of hazardous environments.

For taking an example of this kind of technology, we can consider a micro air vehicle (MAV) as a flying robotic insect such as RoboBee developed at Harvard Microrobotics Laboratory, where a novel fabrication process to create centimeter-scale wings of great complexity using photolithography is introduced (Shang et al. 2009); as resultant wings can be produced with a wide range of desired mechanical and geometric characteristics. It should be borne in mind that mimicking the wing kinematics of biological flight requires examining the potential effects of wing morphology on flight performance.

Miniature robots or nanorobots may be seem like too far; however, very soon, they could be released even into our bodies to maintain and repair them.

5.5.2 Cooperative Robot Teams

The coordination of multiple robots in the execution of cooperative tasks more efficiently and robustly is one of the main challenges facing us today. The actions performed by each team member during each phase of the cooperation must be specified in function of the robot properties, task requirements, and characteristics of the environment. Moreover, the coordination mechanism has to provide flexibility and adaptability. Recently, some researchers have used for the coordinating of real robots a new procedure that consists in a dynamic role assignment (Chaimowicz et al. 2002). In those works, the robots can exchange leadership in a cooperative manipulation task, adapting their coordination patterns in the presence of unexpected events. On the other hand, many biologically inspired systems in order to perform tasks that require distinct roles to be concurrently filled and which cannot be performed by a single robot use a set of well-defined abstractions and techniques for behavior interaction and control, known as port arbitrated behavior (PAB) paradigm. Some studies have demonstrated that the PAB paradigm has reached higher levels of competence than traditional AI methods combined into hybrid systems (Werger and Mataric 2000).

Multirobot teams have been proposed to improve the monitoring and resource utilization when sampling large-scale environmental processes. An example is the COMETS project, which designed and implemented cooperative systems for autonomous environmental perception, including fire detection and monitoring, and terrain mapping, using multiple heterogeneous UAVs (Ollero et al. 2005). To address many of the proposed larger environmental monitoring problems, frameworks for integration of disparate sensing platforms as well as for robot and sensor network interaction and information sharing are necessary.

5.5.3 Smartdevice-Based Robots

Smartphone and tablets offer to robots a combination of sensors, CPU, display, and network connectivity; this is the reason why new robot applications using smart devices are recently emerging constituting a huge wave of innovation. Robotics take advantage of mobile devices (based on Apple's iOS and Google's Android) that can support mobile computing environment. Lately, some remote control methods of a mobile robot using a smartphone with Windows mobile and Bluetooth data communication to support various OS Platforms have been developed (Seo et al. 2011).

Also, some companies have introduced new products using this technology. For instance, iRobot has developed a remote presence prototype robot called Ava, which uses a tablet to control its mobile base, and another example can be Romotive, a little smartphone-powered robot.

5.5.4 High Accuracy 3D Sensing

New 3D sensors provide an effective way of 3D scanning everyday objects, generating libraries that robots would access to know "the real world." Machine intelligence and adaptation require robust real-world sensing. Substantial economic resources are directed toward development of robust real-world sensors and fusion algorithms that enable the robots to fully experience and understand the environment.

High-resolution 3D scanning is essential to improve the performance of object detection, what constitute a critical task to the operation of mobile manipulator robots, especially in homes and in workplaces. Some researchers have demonstrated that augmenting state-of-the-art computer vision techniques with high-resolution 3D information results in higher precision and recall.

Kinect, the Microsoft three-dimensional (3D) sensor, is cheap and easy to use. It has made 3D mapping and motion sensing accessible (low-cost alternative to laser range finders). Kinect2 will feature a higher resolution and frame rate that will allow the device to read lips.

Computational cameras (as Lytro, developed at Stanford) capture both intensity and angle of light, allowing the refocusing of already-snapped pictures and the creation of 3D images (Ng 2006). The enhanced camera samples the total geometric distribution of light passing through the lens in a single exposure.

5.5.5 Cognitive Systems

Cognitive systems have been created to engage the inherent functions of human cognition and increase one's cognitive capabilities. The original basis for cognitive systems was based on theories from psychology and is also part of artificial intelligence (AI) movement. Cognitive processes can come in a variety of forms, such natural or imitation, and can be either conscious or unconscious. Cognitive systems include complex realms of the mind, perception, intelligence, learning as well as the expectations of the artificial mind.

Even the most powerful computers cannot match the human brain. Researchers all over the world are building intelligent cognitive machines that can imitate many behaviors of humans, including thoughts and speech. An example is the cognitive computers used to beat the best human chess players. It expects that these cognitive computers and cognitive systems will incorporate levels of consciousness in the future.

5.5.6 Cloud Robotics

Cloud robotics is a new trend by which robots rely on cloud computing infrastructures to access vast amounts of processing power and data. In this way, robots will benefit from offload compute-intensive tasks such as image processing, voice recognition, or even new skills instantly.

Cloud robotics (Ingebretsen 2011) also refers to the use of Web services in order to make robots more intelligent. This means that robots could leave more complex tasks in the hands of remote servers that would perform computationally heavy operations. Also this means that robots could automatically learn new skills when facing a new situation simply getting out new applications. Therefore, by putting the network to reach, these robots may be more optimal.

Being connected to the cloud will help the robots to collaborate with other machines, smart objects, and humans. Through this collaboration, robots transcend beyond their physical limitations and are more useful and capable as they delegate some of their more specialized tasks to third parties.

Another consequence is that when connected to the cloud and other, robots can automatically report any information when it detects a special situation. Through a service, it is able to create algorithms that detect these situations. These algorithms will be in the cloud, and the robots can use them through a sensor status in real time. This intelligent control or monitoring may take action when it is necessary and inform a human being.

The future of Internet of Things as well as cloud robotics focuses on that everything will be an application, so that there could be a app store for robots functions. At any time, a robot would be able to download an application in order to complete an operation: learning German, prepare a turkey for Christmas, or control a boat for example. Equally it would be useful as a platform where developers could create applications for any type of robot.

The future of robots to operate intelligently and naturally cannot be restricted to physical and individual platforms but focused on a cloud service where they provide an intelligent learning that can be transferred to any platform whether real or virtual and whose resources and information are available at any time.

To give an example, currently Google has a small team creating robot-friendly cloud services. Also, it can be mentioned RoboEarth, an Europe project, whose main goal is to develop a World Wide Web for robots, a giant cloud-enable database where robots can share information about objects, environments, and tasks.

5.5.7 Rapid Prototyping with 3D Printers

Rapid prototyping (RP) can be defined as a group of techniques used to quickly fabricate a scale model of a part or assembly using three-dimensional computer, what allows to figure out whether a design is really useful or not. Through 3D printers, it is possible to create 3D objects from computer-aided design models.

There are endless variations on how to print in three dimensions. Most devices fall into two main categories: extruders and consolidators. An extruder squirts a liquid out of one or more nozzles so that the liquid solidifies, layer by layer, to build up the desired shape. A consolidator spreads a thin layer of some easily removable substance over the build platform, creates bonds at particular spots on that layer, lowers the platform, spreads another thin layer on top of the one that was just processed, and repeats. In the end, the unconsolidated material is removed.

3D printers can be used to create components for the quick and inexpensive development of force sensors, which are essential components in a large range of devices and systems, including robotics. 3D printers will provide inexpensive and easily customized force sensors for robots, improving the performance and high-precision measures through the force information (Kesner and Howe 2011a).

Someday, lots of stuff will be manufactured this way, on demand. Nowadays, 3D printers are used in different environment that even we cannot image and we use daily: aircraft companies that print assembly-jig inserts for holding wing section and other parts in place for drilling and fastening. Other example is the companies that manufacture high-performance cars; they have also begun 3D printing molds for carbon composite panels.

We can find several examples as the MakerBot Industries' Thing-O-Matic 3D printer is a robot that behaves as a rapid prototyping device. This robot has taken out three-dimensional printers of academic and industrial robotics laboratories, allowing people to model and refine designs of digital products for home fabrication.

What all these examples have in common is the high cost of conventional production. So, it makes sense to spend one time hundreds of thousand for 3D printers that can churn out many different plastic shapes with ease.

5.5.8 Development of Verification and Synthesis Techniques

Verification and synthesis are dual areas that share common foundations in how logic functions are stored and manipulated on the computer. Improvements in formal methods, a basic area in computer science, provide great advances in other fields as robotic. In order to ensure that robots will always behave properly, a correct design from the specification is mandatory.

New approaches present the idea that a robot can be modeled as a hybrid system, that is, a system that is defined by both discrete and continuous variables. Researchers try to derive control laws that allow the robot reaches a desired state while avoiding a set of bad states. With respect to the problem of generating correct-by-construction control that assures the robot to achieve a high-level abstract behavior, some works pointed that the use of linear temporal logic (LTL) as the specification formalism can be a successful approach (Molengraft et al. 2011). Though this logic, the truth value of the propositions can change with time, in such way as to avoid the risk of fails about logical formulas that have temporal aspects.

5.6 Conclusions

Like never before, new technological developments can improve our lives. Up until now, robots display an extremely good degree of efficiency in factories but are very incompetent at homes. That is the reason why nowadays most robots work in factories and laboratories; however, in the very next future, a new generation of robots will become in cohabitants very close to us. They are not yet the walking, talking, and intelligent machines of the movies, but ongoing exciting projects promise to deliver more capable autonomous machines in years to come. Adaptation will be one of their main characteristics. Now, we have robots that can adjust their level of autonomy on the fly to achieve the level of control specified by the user, and soon there will be robots that exhibit emotion. Through the development of a variety of machine learning techniques and advanced artificial neural network architectures, robots will be able to incrementally acquire new knowledge from autonomous interactions with the environments, which will form the basis for accomplishing tasks by means their designers did not explicitly implement them. In this sense, we can say that there will not be two robots exactly alike, as they will make their own decisions.

Humanoids will change definitively the way we interact with machines, as they will be designed to interact with people, the interface will be our brains. They will be biological inspirited, and the imitation of the detection capabilities of human skin will be an essential issue. The industrialization of sensor technologies will contribute to the development of sensors more affordable, more variety, more robust and reliable. In order to get a continuous coverage, advanced sensors will extend through the body of the robot. That will constitute the basis for operations of the robot in the real world, which have to ensure the safety, the effectiveness, and efficiency of performance requirements.

New robots are able to operate from a wide variety of platforms, from earth vehicles to submersible robots or even whole colonies of interactive robots. The common element and the main distinguishing feature is that they are prepared to work in unstructured and changing environments where interruptions in communication links, component failures, fast-evolving mission requirements, and similar events frequently happen.

In light of the above, in this chapter, we can say that in years to come, technological developments along with engineers' imagination will create a special synergy from which future robotics will benefit as well as the progress of humankind as a whole.

References

Allen B, Stokey R, Austin T, Forrester N, Goldsborough R, Purcell M, von Alt C (1997) REMUS: a small low cost AUV: system description, field trials, performance results. Proc. MTS/IEEE OCEANS 1997:994–1000

Aucouturier JJ, Ikeuchi K, Hirukawa H, Nakaoka S, Shiratori T, Kudoh S, Kanehiro F, Ogata T, Kozima H, Okuno HG, Michalowski MP, Ogai Y, Ikegami T, Kosuge K, Takeda T, Hirata Y (2008) Cheek to chip: dancing robots and AI's future, IEEE Intell Syst 23(2)

Bingham B, Foley B, Singh H, Camilli R, Delaporta K, Eustice R, Mallios A, Mindell D, Roman C, Sakellariou D (2010) Robotic tools for deep water archaeology: surveying an ancient shipwreck with an autonomous underwater vehicle. J Field Robot 27(6):702–717

Bongard J (2011) Morphological change in machines accelerates. In: Proceedings of the national academy of sciences (PNAS), vol 6. Jan 10

Chaimowicz L, Campos M, Kumar V (2002) Dynamic role assignment for cooperative robots. IEEE Conference on robotics and automation. Washington

Cornblatt M, Sparky Jr (2012), http://sparkyjr.ning.com/. Accessed May 2012

Coxworth B (2011) Ant-Roach illustrates potential for inflatable robots. http://www.gizmag.com/ant-roach-inflatable-robot/20619/. Accessed May 2012

Degani A, Feng S, Choset H, Mason M (2010) Minimalist, dynamic, tube climbing robot. IEEE international conference on robotics and automation 2010

Dunbabin M, Corke P, Buskey G (2004) Low-cost vision-based AUV guidance system for reef navigation. Proc IEEE ICRA 1:7–12

Dunbabin M, Grinham A, and Udy J (2009) An autonomous surface vehicle for water quality monitoring. In: Proceedings of Australasian conference on robotics and automation 2009, Sydney, Australia, p 13

Dunbabin M, Marques L (2012) Robotics for environmental monitoring. IEEE Robot Autom Mag 19(1)

Elkins L, Sellers D and Monach WR (2008) The Autonomous maritime navigation (AMN) project: field tests, autonomous and cooperative behaviors, data fusion, sensors and vehicles. J Field Robot 27

Elston JS, Roadman J, Stachura M, Argrow B, Houston A, Frew EW (2011) The tempest unmanned aircraft system for in situ observations of tornadic supercells: design and VORTEX2 flight results. J Field Robot 28(4):461–483

Gould J, Roemmich D, Wijffels S, Freeland H, Ignaszewsky M, Jianping X, Pouliquen S, Desaubies Y, Send U, Radhakrishnan K, Takeuchi K, Kim K, Danchenkov M, Sutton P, Kind B, Owens B and Riser S (2004) Argo profiling floats bring new era of in situ ocean observations. Eos Trans AGU 85(19):179, 190–191

Griffiths G, Millard N, McPhail S, Stevenson P, Perrett J, Peabody M, Webb A, and Meldrum D (1998) Towards environmental monitoring with the Autosub autonomous underwater vehicle. In: Proceedings of international symposium on underwater technology, p 121–125

Guccione S, Muscato G, Nunnari G, Virk GS, Azad AKM, Semerano A, Ghrissi M, White T and Glazebrook C (2000) Robots for volcanos: the state of the art. In: Proceedings of 3rd international conference on climbing and walking robots (CLAWAR), Madrid, Spain, p 777–788

Guizzo E (2010) DARPA Seeking to revolutionize robotic manipulation. IEEE Spectrum robotics blog. http://spectrum.ieee.org/automaton/robotics/robotics-software/darpa-arm-program. Accessed Jun 2012

Guizzo E (2012) Soft robotics. IEEE Robot Autom 19(1):123–128

Guizzo E, Deyle T (2012) Robotics trends for 2012. IEEE Robot Autom Mag 19(1)

Haapasalo L, Samuels P (2011) Responding to the challenges of instrumental orchestration through physical and virtual robotics. Computers and Education. Elsevier 57(2)

Hirata Y, Komatsuda S, Kosuge K (2008) Fall prevention control of passive intelligent walker based on human model. In: Proceedings of the IEEE/RSJ international conference on intelligent robots and systems, IROS'08

Hitz G, Pomerleau F, Garneau M, Pradalier C, Posch T, Pernthaler J, and Siegwart RY (2012) Autonomous inland water monitoring. IEEE Robot Autom Mag 1:62–72

Ikemoto S, Nishigori Y, Hosoda K (2011) Adaptive Motion of a muscloskeletal robot arm utilizing physical constraint. In: Proceedings of AMAM 2011, pp 93–94

Ingebretsen M (2011) Robotics trends. Thought leader interview: Mario Tremblay: http://www.roboticstrends.com/design_development/article/thought_leader_interview_mario_tremblay. Accessed April 12, 2012

Ishiguro H (2007) Android science: Toward a new cross-interdisciplinary framework. Robotics Research, vol. 28. Springer, Berlin, pp 118–127

Iwata H, Hoshino H, Morita T, Sugano S (2000) Human-humanoid physical interaction realizing force following and task fulfilment. In: Proceedings of the IEEE/RSJ international conference on intelligent robots and systems, IROS'00

Iwata H, Sugano S (2005) Human-robot-contact-state identification based on tactile recognition. IEEE Trans Ind Electron 52(6):1468–1477

Jervis C (2005) Carebots in the Community. British Journal of Healthcare Computing and Information Management 22(8)

Jung S, Kang S, Lee M, Moon I (2007) Design of robotic hand with tendon-driven three fingers. In: Proceedings of International Conference on Control, and Automatic System, Seoul, Korea, pp 83–86

Kemp C, Edsinger A, Torres-Jara E (2007) Challenges for robot manipulation in human environments. IEEE Robot Autom Mag 20

Kesner S, Howe R (2011a) Design principles for rapid prototyping forces sensors using 3-D Printing. IEEE/ASME Trans Mechatron 16(5)

Kesner S, Howe R (2011b) Position control of motion compensation cardiac catheters. IEEE Trans Robot 27(6)

Klinck H, Stelzer R, Jafarmadar K and Mellinger D (2009) AAS endurance: An autonomous acoustic sailboat for marine mammal research. In: Proceedings of International Robotic Sailing Conference (IRSC), Matosinhos, Portugal, pp 43–48

Kuroki Y, Fukushima T, Nagasaka K, Moridaira T, Doi TT, Yamaguchi J (2003) A small biped entertainment robot exploring human–robot interactive applications. IEEE International Symposium on robot and human interactive communication, RO-MAN'03

Leonard N, Paley D, Davis R, Fratantoni D, Lejien F, Zhang F (2011) Coordinated control of an underwater glider fleet in an adaptive ocean sampling field experiment in Monterey Bay. J Field Robot 27(6):718–740

Lim H, Tanie K (1999) Collision-tolerant control of human-friendly robot with viscoelastic trunk. IEEE Trans Mechatron 4(4):417–427

Lin PH, Lee CS (2008) The eyewall-penetration reconnaissance observation of typhoon longwang with unmanned aerial vehicle, aerosonde. J Atmos Ocean Technol 25(1):15–25

Markoff J (2010) Google cars drive themselves, in traffic. The New York Times

McMaster S (2012) Idaho national laboratory. https://inlportal.inl.gov/portal/server.pt/community/robotics_and_intelligence_systems/455. Accessed May 2012

Mizuuchi I, Nakanishi Y, Sodeyama Y, Namiki Y, Nishino T, Muramatsy N, Urata J, Hongo K, Yoshikai T, Inaba M (2007) An advanced musculoskeletal humanoid kojiro. 7th International conference on humanoid robots

Molengraft R, Beetz M, Fukuda T (2011) Robot challenges: toward development of verification and synthesis techniques. IEEE Robot Autom Mag 18(4)

Mukai T, Onishi M, Odashima T, Hirano S, Luo Z (2008) Development of the tactile sensor system of a human-interactive robot RI-MAN. IEEE Trans Rob 24(2):505–512

Ng R (2006) Digital light field photography. Dissertation. Department of computer science. Stanford University

Nishio S, Ishiguro H, Hagita N (2007) Geminoid: teleoperated android of an existing person. In: de Pina Filho AC (ed) Humanoid robots: new developments, I-Tech, Vienna, Austria

Nuria M, Oliver, BR (2000) A Bayesian computer vision system for modeling human interactions. IEEE Trans Pattern Anal Mach Intell 22(8)

Odashima T, Onishi M, Tahara K, Mukai T, Hirano S, Luo Z, Hosoe S (2007) Development and evaluation of a human-interactive robot platform 'RI-MAN'. J Robot Soc Jpn 25(4):554–565 (in Japanese)

Odashima T, Onishi M, Tahara K, Takagi K, Asano F, Kato Y, Nakashima H, Kobayashi Y, Luo ZW, Mukai T and Hosoe S (2006) A soft human-interactive robot—RI-MAN—Video. In: Proceedings of IEEE/RSJ international conference on intelligent robots and systems

Ohmura Y, Kuniyoshi Y (2007) Humanoid robot which can lift a 30 kg box by whole body contact and tactile feedback. In: Proceedings of the IEEE/RSJ international conference on intelligent robots and systems, IROS'07

Oliver N (2000) Towards perceptual intelligence: statistical modeling of human individual and interactive behaviors. PhD thesis, Massachusetts Institute of Technology (MIT), Media Lab, Cambridge, Mass

Oliver N, Rosario B, and Pentland A (1999) A Bayesian computer vision system for modeling human interactions. In: Proceedings of international conference on vision systems

Ollero A, Lacroix S, Merino L, Gancet J, Wiklund J, Remuss V, Perez IV, Gutierrez LG, Viegas DX, Benitez MAG, Mallet A, Alami R, Chatila R, Hommel G, Lechuga FJC, Arrue BC, Ferruz J, Martinez-De Dios JR, Caballero F (2005) Multiple eyes in the skies: architecture and perception issues in the COMETS unmanned air vehicles project. IEEE Robot Autom Mag 12(2):46–57

Palli G, Borghesan G, Melchiorri C (2009) Tendon-based transmissions systems for robotic devices: models and control algorithms. In: Proceedings of international conference on robotics and automation. Kobe, Japan, pp 4063–4068

Palli G, Borghesan G, Melchiorri C (2010) Friction and visco-elasticity effects in tendon-based transmission systems. In: Proceedings of International Conference on Robotics and Automation. Anchorange, AK, pp 3890–3895

Palli G, Borghesan G, Melchiorri C (2012) Modeling, identification and control of tendon-based actuation systems. IEEE Trans Rob 28(2):277–290

Pfeifer R (2010) "Soft robotics": self-organization, embodiment, and biological inspiration. Artificial Intelligence Laboratory, University of Zurich, Switzerland. http://www.iiitb.ac.in/uploads/PfeiferAbstractShortBio.pdf. Accessed May 2012

Pfeifer R. ECCE project (2011). http://eccerobot.org. Accessed Jun 2012

Pferifer R, Lungarella M, Iida F (2007) Self-organization, embodiment and biologically inspired robotics. Science 318(5853):1088–1093

Ramana M, Ramanathan V, Kim D, Roberts G, Corrigan C (2007) Albedo, atmospheric solar absorption and heating rate measurements with stacked UAVs. Quart J Royal Meteorol Soc 133(629):1913–1931

Rynne P, von Ellenrieder K (2009) Unmanned autonomous sailing: current status and future role in sustained ocean observations. Marine Technol Soc J 43(1):21–30

Schmidt PA, Maél E, Würtz RP (2006) A sensor for dynamic tactile information with applications in human-robot interaction and object exploration. Robot Auton Syst 54:1005–1014

Scott M (2012) JamBots: soft robots based on particle jamming, like this hexapod from iRobot. http://onlineallthenews.blogspot.com.es/2012/01/jambots-soft-robots-based-on-particle.html. Accessed May 2012

Seo Y, Kwak S, Yang T (2011) Mobile robot control using smart phone and its performance evaluation. Communications in computer and information science. Advanced communication and networking. Third international conference, vol 199. ACN

Shang J, Combes S, Finio B, Wood R (2009) Artificial insect wings of diverse morphology for flapping-wing micro air vehicles. Bioinspiration Biomimetics 4(3)

Shibata T, Tanie K (2001) Physical and affective interaction between human and mental commit robot. In: Proceedings of the ieee international conference on robotics and automation, ICRA'01

Smith R, Das J, Heidarsson H, Pereira A, Arrichiello F, Cetinic I, Darjany L, Garneau ME, Howard M, Oberg C, Ragan M, Seubert E, Smith E, Stauer B, Schnetzer A, Toro-Farmer G, Caron D, Jones B and Sukhatme GS (2010a) The USC Center for Integrated Networked Aquatic PlatformS (CINAPS): observing and monitoring the Southern California Bight. IEEE Robot Autom Mag 17:20–30 (Special Issue on Marine Robotic Systems)

Smith RN, Chao Y, Li PP, Caron DA, Jones BH and Sukhatme GS (2010b) Planning and implementing trajectories for autonomous underwater vehicles to track evolving ocean processes based on predictions from a regional ocean model. Int J Robot 26(12). http://cres.usc.edu/cgi-bin/print'pub'details.pl?pubid=646. Accessed May 2012

Squyres SW, Arvidson RE, Bell JF, Brückner J, Cabrol NA, Calvin W, Carr MH, Christensen PR, Clark BC, Crumpler L, Des Marais DJ, d'Uston C, Economou T, Farmer J, Farrand W, Folkner W, Golombek M, Gorevan S, Grant JA, Greeley R, Grotzinger J, Haskin L, Herkenhoff KE, Hviid S, Johnson J, Klingelhöfer G, Knoll A, Landis G, Lemmon M, Li R, Madsen MB, Malin MC, McLennan SM, McSween HY, Ming DW, Moersch J, Morris RV, Parker T, Rice JW, Richter L, Rieder R, Sims M, Smith M, Smith P, Soderblom LA, Sullican R, Wänke H, Wdowiak T, Wolff M and Yen A (2004) The spirit Rover's Athena science investigation at Gusev Crater, Mars, Science 305(5685):794–799

Stealey M, Singh A, Batalin M, Jordan B and Kaiser W (2008) NIMSAQ: a novel system for autonomous sensing of aquatic environments. In: Proceedings of IEEE ICRA, Pasadena, CA, pp 621–628

Stiehl WD, Lieberman J, Breazeal C, Basel L, Lalla L, Wolf M (2005) The design of the Huggable: a therapeutic robotic companion for relational, affective touch. In: Proceedings of the AAAI fall symposium on caring machines: AI in Eldercare

Takamuku S, Fukuda A, Hosoda K (2008) Repetitive grasping with anthropomorphic skin-covered hand enables robust haptic recognition. In: Proceedings of IEEE/RSJ international conference on intelligent robots and systems, ThAT13.5

Takeda T, Hirata Y, Kosuge K (2007) HMM-based error recovery of dance step selection for dance partner robot. In: Proceedings of the IEEE international conference on robotics and automation, ICRA'07

Weekly K, Anderson L, Tinka A, and Bayen A (2011) Autonomous river navigation using the Hamilton-Jacobi framework for underactuated vehicles. In: Proceedings of IEEE conference on robotics and automation, Shanghai, China, pp 828–833

Werger B, Mataric M (2000) Broadcast of local eligibility: behavior-based control for strongly cooperative robot teams. In: Proceedings of the fourth international conference on Autonomous agents

Chapter 6
Robotics in Alternative Energy

Raquel Dormido Canto and Natividad Duro Carralero

6.1 Introduction

Energy is a fundamental task in the modern economy and society, and access to secure, reliable, and competitively priced energy has been for years a cornerstone in the global economic and social development.

Even though fossil fuels still play a major role in global energy needs, with its rising costs and policies to reduce greenhouse gas emissions, renewable and alternative energies are being looked into.

Achieving these clean energy ambitions will require the global community to successfully develop, adapt, commercialize, and deploy new technologies and processes across a range of energy applications. This will include not just energy generation but also transport energy storage, grid management, building design, and more efficient end-use technologies.

Actually, nations and manufacturers are turning their scientific research to alternative sources of power. Large-scale investment must be delivered to meet the growing demand for energy and improve energy productivity. Wind, solar, and biological continue their remarkable growth as alternative energies poised to supplant coal and oil. Nevertheless, until now, the cost per megawatt is higher than conventional sources.

In this context, the importance of robotics in the development and use of alternative energy is clear. It can be read robots as "critical to survival of alternative energy industry." They play a leading role in making ever-changing alternative energy more competitive. Robotics offers intelligent solutions in the manufacturing processes. Robotics in combination with control helps manufacturers to reduce costs, improve quality, and increase the productivity.

R. Dormido Canto (✉) · N. Duro Carralero
Dpto. Informatica y Automatica, UNED, 28040 Madrid, Spain
e-mail: raquel@dia.uned.es

N. Duro Carralero
e-mail: nduro@dia.uned.es

A. López Peláez (ed.), *The Robotics Divide*,
DOI: 10.1007/978-1-4471-5358-0_6, © Springer-Verlag London 2014

For instance, solar companies are worried about the growing demand and the need to drive down the costs per watt requires that their automation process delivers high yields, high volume, quick change in capacity, flexibility to adapt to part variations, and consistent product inspection throughout the process. Robotics is a good partner for solar manufacturing: robots can integrate vision and inspection to cover different solar manufacturing needs. For example, one of the challenges with solar power is the amount of space panels take compared to other energy sources. Pointing solar panels toward the sun with trackers is a common way to get more energy from panels, particularly at large-scale farms. Robotic systems have been developed for tracking the sun along two dimensions, but at lower cost than traditional two-axis trackers. Robots are also used in the installation of large photovoltaic panel during the construction of a solar plant is done using a robotic arm. Another example is the robotic solutions incorporated in the construction and finishing of wind turbines. Fast and flexible robots allow not only to increase production but also to reach a better precision than with manual methods.

On the other hand, alternative energies can be used by robots as own energy source. Mobile robots have demonstrated their versatility in a wide range of applications. However, they have their limitations because of their reliance on traditional energy source which, by their very nature, cannot provide a convenient source of energy for all applications. For example, there are those robots which are reliant on electrical power supply, and they need to be plugged into an electrical power source all the time in order to be operational, which tends to limit their range of operation. In addition, electrical power supply may not always be available in certain situations, as for example, in disaster settings where such type of power is almost invariably interrupted or may be altogether unavailable due to damage to infrastructure, thereby making an electrically operated robot not a very viable option. The other alternative source of energy that could be employed to power a mobile robot is battery power, but even this option has its limitations, as batteries cannot provide a durable source of power over extended periods of time normally required for sustained operations which are necessary in search and rescue efforts. Yet another alternative energy source that could be employed to power a mobile robot is solar power. Robots can carry solar panels on it that help recharge its onboard battery. Once fully charge the robot can be used to the end, it has been designed. Nevertheless, this energy source has limitations. A solar-powered robot would be dependent on a constant light source to be able to function, and therefore, would not be viable for use in adverse light conditions, more especially in search and rescue operations where it would be needed most to seek out victims trapped underground or under rubble away from light. This limitation could be dealt if, for example, it is investigated the viability of a solar thermal energy storage system to overcome this time mismatch between solar energy availability and demand.

Anyway, the need for new investment in robotics and alternative energies is particularly critical to strike an appropriate balance in delivering energy security, facilitating economic development, and meeting clean goals.

This chapter is focused on three viable forms of alternative energy: solar power, wind power, and biological energy. Several aspects of these alternative forms of energy are described. The role of robotics in each case is also analyzed.

6.2 Solar Energy

Solar energy, radiant light, and heat from the sun, has been harnessed by humans since ancient times, using a range of ever-evolving technologies. Solar energy technologies include many different applications, which can make considerable contributions to solving some of the most urgent problems the world now faces (OECD/IEA 2011).

Solar energy is the energy obtained by capturing the electromagnetic radiation and the heat emitted by the sun. The amount of sun reaching the earth is higher than the total present need for energy on Earth. Approximately 30 % of the sun's radiation is reflected back to space while the rest is absorbed by clouds, oceans, and land masses.

Earth's land surface, oceans, and atmosphere absorb solar radiation, and this raises their temperature. Warm air containing evaporated water from the oceans rises, causing atmospheric circulation or convection. When the air reaches a high altitude, where the temperature is low, water vapor condenses into clouds, which rain onto the Earth's surface, completing the water cycle. The latent heat of water condensation amplifies convection, producing atmospheric phenomena such as wind, cyclones, and anti-cyclones. Sunlight absorbed by the oceans and land masses keeps the surface at an average temperature of 14 °C (Somerville 2007). By photosynthesis, green plants convert solar energy into chemical energy, which produces food, wood, and the biomass from which fossil fuels are derived (Vermass 2007).

However, although the Earth receives a lot of energy from the sun, it is only possible convert 10–15 % of this sun energy to usable energy (Pasqualetti and Miller 1984). This is clearly not as efficient as coal power, 40 % energy efficiency or wind power, 20 % energy efficiency. Furthermore, while coal energy and wind energy can be used for all day long, solar energy can only use during the day when there is sunlight. This situation can only be worse during winter.

Despite this, solar energy is deemed to be the future energy. Besides that, solar energy is the cleanest energy because of it is totally pollution-free, and for that, it is considered as a perfect solution for the energy needs in the world.

This energy is clean and easy to maintain. It is a so-called renewable energy and in particularly belongs to the group known as clean energy or green energy, but at the end of its life, for example, the solar panels used in thermal plants can be a difficult pollutant recyclable waste today.

In 2011, the International Energy Agency said that "the development of affordable, inexhaustible and clean solar energy technologies will have huge longer-term benefits. It will increase countries' energy security through reliance on an indigenous, inexhaustible and mostly import-independent resource, enhance sustainability, reduce pollution, lower the costs of mitigating climate change, and keep fossil fuel prices lower than otherwise. These advantages are global. Hence the additional costs of the incentives for early deployment should be considered learning investments; they must be wisely spent and need to be widely shared" (OECD/IEA 2011).

Solar energy represents a very abundant and inexhaustible energy resource to mankind, relatively well-spread over the globe. Its availability is greater in warm and sunny countries. These countries will likely contain about seven billion inhabitants by 2050 versus two billion in cold and temperate countries (including most of Europe, Russia, and parts of China and the United States of America).

The cost of the production of the solar energy depends on the used solar energy technology. There are many different forms of solar energy production as, solar heating, solar photovoltaic, solar thermal electricity, and solar architecture, with different advantages and disadvantages.

The costs of solar energy have been falling rapidly and are entering new areas of competitiveness. For example, solar thermal electricity and solar photovoltaic electricity are every time more competitive against oil-fueled electricity generation. However, in the most markets, solar electricity is not yet able to compete without specific incentives.

6.2.1 Applications of Solar Energy

Solar technologies are broadly characterized as either passive solar or active depending on the way they capture, convert, and distribute solar energy. **Active** *solar techniques* include, for example, the use of photovoltaic panels, solar thermal collectors, pumps, and fans to harness the energy. *Passive solar techniques* include orienting a building to the sun, selecting materials with favorable thermal mass or light dispersing properties, and designing spaces that naturally circulate air. These features must be adequate to the local climate and environment, in order to obtain the best comfortable temperature range.

Active solar technologies increase the supply of energy and are considered supply-side technologies, while passive solar technologies reduce the need for alternate resources and are generally considered demand-side technologies (Philibert 2005).

6.2.1.1 Solar Energy and Agriculture

Nowadays, the agriculture uses the capture of solar energy in order to optimize the productivity of plants. Techniques such as timed planting cycles, tailored row orientation, staggered heights between rows, and the mixing of plant varieties can improve crop yields (Jeffrey 1999; Kaul et al. 2001). The solar energy is very important to the plants.

Applications of solar energy in agriculture aside from growing crops include pumping water and drying crops. More recently, the technology has been embraced to use the energy generated by solar panels to power grape presses. But the most principal application is in the greenhouse.

A *greenhouse* converts solar light to heat, enabling year-round production and the growth (in enclosed environments) of specialty crops and other plants not naturally suited to the local climate.

A greenhouse is a structure with different types of covering materials, such as a glass or plastic roof and frequently glass or plastic walls. It heats up because incoming visible solar radiation from the sun and this radiation is absorbed by plants, soil, and other things inside the building. Air warmed by the heat from hot interior surfaces is retained in the building by the roof and wall (see the Fig. 6.1).

Thus, the primary heating mechanism of a greenhouse is convection. This can be demonstrated by opening a small window near the roof of a greenhouse: the temperature drops considerably. This used principle is the basis of the auto ventilation automatic cooling system.

6.2.1.2 Solar Lighting

In the twentieth century, artificial lighting became the main source of interior illumination but delighting techniques and hybrid solar lighting solutions are ways to reduce energy consumption.

Fig. 6.1 Greenhouse (this is a figure from the Wikimedia Commons)

Day lighting systems collect and distribute sunlight to provide interior illumination. This passive technology directly offsets energy use by replacing artificial lighting and indirectly offsets non-solar energy use by reducing the need for air-conditioning. Day lighting design implies careful selection of window types, sizes, and orientation. Individual features include saw tooth roofs, clerestory windows, light shelves, skylights, and light tubes. They may be incorporated into existing structures, but are most effective when integrated into a solar design package that accounts for factors such as glare, heat flux, and time-of-use. When delighting features are properly implemented they can reduce lighting-related energy requirements by 25 % (Apte et al. 2003).

Hybrid solar lighting is an active solar method of providing interior illumination. These systems collect sunlight using focusing mirrors that track the sun and use optical fibers to transmit it inside the building to supplement conventional lighting. Applications of these systems are able to transmit 50 % of the direct sunlight received (Muhs and Oak Redge National Lab 2000).

6.2.1.3 Solar Thermal for Water Heating

Solar hot water systems use sunlight to heat water. In low geographical latitudes, it is possible to generate until 70 % of the domestic hot water. The most common types of solar water heaters are *evacuated tube collectors* (44 %), *glazed flat plate collectors* (34 %) generally used for domestic hot water, and *unglazed plastic collectors* (21 %) used mainly to heat swimming pools (see the Fig. 6.2) (Weiss et al. 2005).

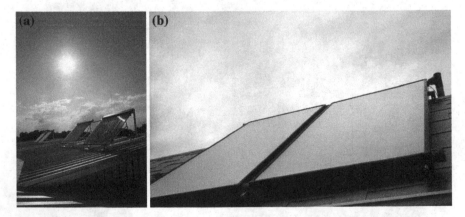

Fig. 6.2 Water heating with solar thermal. **a** Evacuated tube collectors. **b** Glazed Flat plates collectors (this is a figure from the Wikimedia Commons)

6.2.1.4 Solar Thermal for Heating, Cooling, and Ventilation

The *heating, ventilation, and air-conditioning systems* account an important part of the energy used in commercial buildings and in residential buildings. Solar heating, cooling, and ventilation technologies can be used to offset a portion of this energy. Thermal mass is any material that can be used to store heat from the sun in the case of solar energy, as for example, stone, cement, and water. They have been used in arid climates, or warm temperate regions, to keep buildings cool by absorbing solar energy during the day and radiating stored heat to the cooler atmosphere at night (see the Fig. 6.3).

A *solar chimney* is a passive solar ventilation system composed of a vertical shaft connecting the interior and exterior of a building. As the chimney warms, the air inside is heated causing an updraft that pulls air through the building. Performance can be improved by using glazing and thermal mass materials (Bright 1977) in a way that mimics greenhouses.

Fig. 6.3 Solar house (this is a figure from the Wikimedia Commons)

6.2.1.5 Solar Thermal for Water Treatment

Solar distillation can be used to make saline or brackish water potable. *Solar water disinfection* involves exposing water-filled plastic bottles to sunlight for several hours. Exposure times vary depending on weather and climate from a minimum of six hours to two days during fully overcast conditions.

6.2.1.6 Solar Thermal for Cooking

Solar cookers use sunlight for cooking, drying, and pasteurization. They can be grouped into three broad categories: *box cookers, panel cookers,* and *reflector cookers* (Anderson and Palkovic 1994).

The simplest solar cooker is the box cooker. A basic box cooker consists of an insulated container with a transparent lid. It can be used effectively with partially overcast skies and will typically reach temperatures of 90–150 °C.

Panel cookers use a reflective panel to direct sunlight onto an insulated container and reach temperatures comparable to box cookers.

Reflector cookers use various concentrating geometries (dish, trough, Fresnel mirrors) to focus light on a cooking container. These cookers reach temperatures of 315 °C and above but require direct light to function properly and must be repositioned to track the sun (see the Fig. 6.4).

Fig. 6.4 Cooking with solar thermal. **a** Box cooker. **b** Reflector cooker (this is a figure from the Wikimedia Commons)

6.2.1.7 Solar Power

Solar power is the conversion of sunlight into electricity, either directly using photovoltaic panels, or indirectly using concentrated solar power.

Concentrated Solar Power

Concentrating solar power systems use lenses or mirrors and tracking systems to focus a large area of sunlight into a small beam. The concentrated heat is then used as a heat source for a conventional power plant. A wide range of concentrating technologies exists, but the most developed are the parabolic trough (See the Fig. 6.5).

Various techniques are used to track the sun and focus light. In all of these systems, a working fluid is heated by the concentrated sunlight and is then used for power generation or energy storage (Martin et al. 2005).

Fig. 6.5 Parabolic through (this photograph is from Warren Gretz, NREL 04511)

Photovoltaic Panel

Photovoltaic panel is a method of generating electrical power by converting solar radiation into direct current electricity using semiconductors that exhibit the photovoltaic effect (see the Fig. 6.6).

Photovoltaic power generation employs solar panels composed of a number of cells containing a photovoltaic material (Mark 2009). Photovoltaic panels are best known as a method for generating electric power by using solar cells to convert energy from the sun into a flow of electrons.

Photovoltaic panel is one of the fastest-growing power generation technologies in the world. Driven by advances in technology and increases in manufacturing scale and sophistication, the cost of photovoltaic has declined steadily.

Net metering and financial incentives, such as preferential feed-in tariffs for solar-generated electricity have supported solar photovoltaic panel installations in many countries. Some representative advantages of the photovoltaic panels are as follows:

- Photovoltaic panel systems can be designed for a variety of applications and operational requirements and can be used for either centralized or distributed power generation.

Fig. 6.6 Photovoltaic panels (this photograph is from SunPower, NREL 13823)

- Photovoltaic panel systems have no moving parts and are modular, easily expandable, and even transportable in some cases.
- Energy independence and environmental compatibility are two attractive features of photovoltaic panel systems.
- The fuel (sunlight) is free, and no noise or pollution is created from operating photovoltaic panel systems.
- Photovoltaic panel systems that are well designed and properly installed require minimal maintenance and have long service lifetimes.

In the other way, some representative disadvantages of the photovoltaic panels are as follows:

- High cost of photovoltaic panel modules and equipment (as compared to conventional energy sources) is the primary limiting factor. Consequently, the economic value of photovoltaic panel systems is realized over many years.
- In some cases, the surface area requirements for photovoltaic panel arrays may be a limiting factor.
- Due to the diffuse nature of sunlight and the existing sunlight to electrical energy conversion efficiencies of photovoltaic devices, special surface area requirements are needed.
- In climates with many cloudy days, power output is reduced from its full potential, which means that your initial investment takes longer to pay back.

6.2.1.8 Solar Chemical

Solar chemical processes use solar energy to drive chemical reactions. These processes offset energy that would otherwise come from a fossil fuel source and can also convert solar energy into storable and transportable fuels. Solar-induced chemical reactions can be divided into *thermochemical* or *photochemical* (Bolton 1997).

A variety of fuels can be produced by artificial photosynthesis (Wasielewski 1992). The multielectron catalytic chemistry involved in making carbon-based fuels (such as methanol) from reduction of carbon dioxide is challenging. A feasible alternative is hydrogen production from protons, though use of water as the source of electrons (as plants do) requires mastering the multielectron oxidation of two water molecules to molecular oxygen (Hammarstrom and Hammes-Schiffer 2009).

Hydrogen production technologies are a very significant area of solar chemical research since the 1970s. Aside from electrolysis driven by photovoltaic or photochemical cells, several thermochemical processes have also been explored.

Other alternative is the *photoelectrochemical cells*, which consist of a semi-conductor, typically titanium dioxide, immersed in an electrolyte. When the semiconductor is illuminated, an electrical potential develops. There are two types of photoelectrochemical cells: photoelectric cells that convert light into electricity and photochemical cells that use light to drive chemical reactions such as electrolysis (Bolton 1997).

6.2.1.9 Solar Vehicles

A *solar vehicle* is an electric vehicle powered completely or significantly by direct solar energy. Usually, photovoltaic cells contained in solar panels convert the sun's energy directly into electric energy. Solar power may be also used to provide power for communications or controls or other auxiliary functions.

Solar vehicles are not sold as practical day-to-day transportation devices at present, but are primarily demonstration vehicles and engineering exercises. However, indirectly solar-charged vehicles are widespread and solar boats are available commercially (see the Fig. 6.7).

Instead of simply listing headings of different levels, we recommend that every heading is followed by at least a short passage of text.

Fig. 6.7 Solar vehicles. **a** Solar car. **b** Solar boat (this is a figure from the Wikimedia Commons)

6.2.2 Control and Robotics in Solar Energy

Several variables, such as the sun's position and atmospheric conditions, can affect the amount of solar energy you can harness. To make the most of the solar energy available, you can use a control system to track the path of the sun while monitoring important data such as ambient conditions and power quality.

Currently, most large systems incorporate single-axis trackers. Single-axis tracking that follows the azimuth of the sun in order to increase efficiency of the system. In some systems, solar panels are made to face the sun directly (see the Fig. 6.8). Other systems consist of mirrors that concentrate the sun's rays by reflecting them onto a central solar collector.

Sun tracking has two benefits: to maximize the amount of energy that you can harness and to distribute the solar flux evenly on a solar collector's surface, thus ensuring a healthier system.

A sun-tracking system can be active or passive. Active tracking signifies that a light-intensity sensor provides solar irradiance feedback for the system to optimize the position of the solar collector.

Passive tracking uses the predetermined known path of the sun to control the direction of the solar collector. Astronomers have provided algorithms that determine the position of the sun based on date, time, and their location on the earth. These algorithms take into account the rotation of the earth and revolution

Fig. 6.8 Solar plant (this photograph is from Hugh Reilly, NREL 02186)

around the sun (seasonal changes) for different longitudes and latitudes and can provide the azimuth and elevation of the sun.

The main components of a *control tracking system* are as follows:

- Sun-tracking algorithm: This algorithm calculates the azimuth and elevation of the sun and outputs the direction that the solar panel or reflector should face. It can either calculate the sun's position using real-time light-intensity readings or using predetermined knowledge of the sun's movement. It is controlled the direction of the panel or mirror.
- Motion control system: In the motion control portion of the system, a motor is driven to rotate the solar panel while an encoder is used to read the current position of the panel. These components are used to ensure that the panel is facing the desired azimuth and elevation.
- Power quality monitoring system: Keeping track of the current, voltage, and power generated by a solar cell is important to monitor efficiency and system health. When you use a solar panel for sun tracking, you can integrate power monitoring into the system. With reflectors, this is not necessary because the power is not harnessed at this location. You can also use these current and voltage readings to perform maximum power point tracking.
- Ambient condition sensing system: Because ambient conditions affect the operating point and efficiency of solar power harvesting, you need to acquire and log readings from sensors that measure light intensity (solar radiance), temperature, humidity, and so on. You also need pyranometers that read light intensity if you are directly using this value in the sun-tracking algorithm.

On the other hand, *industrial robots* effectively accommodate the rapid (and constant) evolution of solar energy technology. Simulation software makes possible for solar energy system companies to change products and processes quickly and seamlessly.

Solar panel manufacturers increasingly rely on robots for a host of material handling applications. Industrial robots provide ideal solutions for precision-driven assembly and heavy palletizing needs, while still assisting with inspection and quality control jobs. Robotic vision eliminates the need for complex elements and plays a key role in many solar energy manufacturing applications, from locating and inspecting parts, to guiding robots.

There are two types of relevant robots used in solar plant:

- *Robotic Cell and Wafer Handling*: Solar cells and silicon wafers are delicate items that require delicate handling solutions. Robots are capable of assembling these solar energy system components with greater gentleness and precision than is possible manually. They maintain consistent production speeds and repeatability necessary for solar cell and silicon wafer assembly/handling.
- *Robots for Solar Panel Assembly* (see the Fig. 6.9): Manufacturers benefit from using robots when it comes to constructing solar panels, which are often very

Fig. 6.9 Robot for solar panel assembly (this photograph is from Pat Corkery, NREL 17161)

complex in design. Once again, consistency and delicate handling are very important. Every component must be carefully aligned. Robots are also very useful for handling heavy glass panels and other solar panel components with care.

6.3 Wind Energy

The conversion of wind energy to various other useful forms, like electricity, is known as wind power. Wind energy is converted into these forms using wind turbines which convert the kinetic energy in the wind into mechanical power. This mechanical power can be used for specific tasks (such as grinding grain or pumping water) or can be converted into electricity using a generator.

The first use of wind energy was through windmills. Windmills had engines used to produce energy using wind. This energy was usually used in rural and agricultural areas for grinding, pumping, hammering, and various farm needs. Even today, wind energy is used in large-scale wind farms (see the Fig. 6.10) to provide electricity to rural areas and other far-reaching locations.

Wind energy is being used extensively in areas like Denmark, Germany, Spain, India, and in some areas of the United States of America. It is one of the largest forms of green energy used in the world today. Wind energy is highly practical in places where the wind speed is 10 mph.

Fig. 6.10 Wind farms have sprung up in recent year (this is a photograph from Iberdrola Renewables, Inc/NREL 15177)

6.3.1 Wind Power: Advantages and Disadvantages

Wind energy offers many advantages, which explains why it is one of the fastest-growing energy sources in the world. *Main advantages* of wind energy are listed below (U.S. Department of Energy 2011):

- Wind energy is renewable.
- It is widely distributed, cheap, and also helps in reducing toxic gas emissions. It is getting cheaper to produce wind energy. Wind energy may soon be the cheapest way to produce energy on a large-scale.
- The cost of producing wind energy is coming down continuously.
- Along with economy, wind energy is also said to diminish the greenhouse effect. Wind energy generates no pollution. It is friendly to the surrounding environment, as no fossil fuels are burnt to generate electricity from wind energy.
- Wind energy could be readily available around the globe, and therefore, there would be no need of dependence for energy for any country.
- Wind energy may be the answer to the globe's question of energy in the face of the rising petroleum and gas prices.

Although wind energy is one of the renewable sources of energy, there are certain constraints which pose a challenge in using it to its full potential. Some *disadvantages* for wind energy are the following (US Department of Energy 2011):

- Wind turbines generally produce less electricity than the average fossil fueled power station, requiring multiple wind turbines to be built in order to make an impact. A large number of turbines have to be built to generate a proper amount of wind energy.

- Many potential wind farms, places where wind energy can be produced on a large-scale, are far away from places for which wind energy is best suited. Therefore, the economical nature of wind energy may take a beating in terms of costs of new substations and transmission lines.
- The noise pollution from commercial wind turbines is sometimes similar to a small jet engine. This is fine if you live miles away, where you will hardly notice the noise, but what if you live within a few hundred meters of a turbine? This is a major disadvantage.
- Wind turbine construction can be very expensive and costly to surrounding wildlife during the build process. In this sense, protests usually confront any proposed wind farm development. Wind turbines typically cover vast expanses of landscapes, and that has received much public discontent (University of Michigan). Typically, land is viewed as sacred, has some esthetic value, or is under some political restrictions can lead to problems with the implementation of such projects—simply because turbines take up a lot of space and are ugly. People feel the countryside should be left intact for everyone enjoy its beauty.
- The amount of wind supplied to a place and the amount of energy produced from it will depend on various factors like wind speeds and the turbine characteristics. Some critics also wonder whether wind energy can be used in areas of high demand.

6.3.2 Control and Robotics in Wind Energy

In order to understand the role of control and robotics in wind energy, a general overview of the wind turbine operation is presented (National Instruments 2008).

Simply stated, a wind turbine works the opposite of a fan. Instead of using electricity to make wind, like a fan, wind turbines use wind to make electricity. A wind turbine converts the kinetic energy from the wind into mechanical energy. This mechanical energy is then converted into electricity that is sent to a power grid. The turbine components responsible for these energy conversions are the rotor and the generator.

Figure 6.11 shows the major components of a wind turbine: gearbox, generator, hub, rotor, low-speed shaft, high-speed shaft, and the main bearing. The components are all housed together in a structure called the nacelle.

The rotor is the area of the turbine that consists of both the turbine hub and blades. As wind strikes the turbine's blades, the hub rotates due to aerodynamic forces. This rotation is then sent through the transmission system to decrease the revolutions per minute. The transmission system consists of the main bearing, high-speed shaft, gearbox, and low-speed shaft. The ratio of the gearbox determines the rotation division and the rotation speed that the generator sees. For example, if the ratio of the gearbox is N to 1, then the generator sees the rotor speed divided by N. This rotation is finally sent to the generator for mechanical-to-electrical conversion.

Fig. 6.11 Components of a
wind turbine, courtesy of
National Instruments (*NI*)

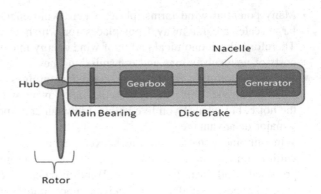

Wind turbine control is necessary to ensure low maintenance costs and efficient performance. The control system also guarantees safe operation, optimizes power output, and ensures long structural life.

Different control methods to either optimize or limit power output can be used. Turbine rotational speed and the generator speed are two key areas that must be controlled for power limitation and optimization. Different techniques used in these areas to control a turbine are briefly described:

- *Blade angle adjustment*: it is also called pitch control (see the Fig. 6.12). The purpose of pitch control is to maintain the optimum blade angle to achieve certain rotor speeds or power output.

Fig. 6.12 Pitch adjustment,
courtesy of National
Instruments (*NI*)

- *Turbine rotation adjustment*: it is also called yaw control (see the Fig. 6.13). Yaw refers to the rotation of the entire wind turbine in the horizontal axis. Yaw control ensures that the turbine is constantly facing into the wind to maximize the effective rotor area and, as a result, power.
- *Speed control*: it deals with the electrical subsystem. This dynamic control is achieved with power electronics, or, more specifically, electronic converters that are coupled to the generator. The two types of generator controls are stator and rotor. The stator and rotor are the stationary and non-stationary parts of a generator, respectively. In each case, the stator or rotor is disconnected from the grid to change the synchronous speed of the generator independently of the voltage or frequency of the grid. Controlling the synchronous generator speed is the most effective way to optimize maximum power output at low wind speeds.

As mentioned before, the main goal of any control method (pitch, yaw, rotational speed control) is to optimize or limit the power extracted from the wind. Wind turbine control is essential for optimal performance, safe operation, and structural stability.

It is worth to note that a common used tool by the control designer is the wind turbine simulators, which provides researchers with a controlled test environment for wind turbine generators, inverters, and systems operations, resulting in improvement research productivity. These simulators provide an infrastructure for the development of advanced control methodologies to improve aspects of system performance such as maximum power extraction from wind or solar sources (Neammanee et al. 2007; Abbas et al. 2010).

Fig. 6.13 Yaw adjustment, courtesy of National Instruments (*NI*)

On the other hand, due to the scale of wind turbine blades and tower pose, industrial robots have proved invaluable tools for nearly every aspect of wind turbine manufacturing. Wind energy system manufacturers incorporate robotic solutions in both the construction and finishing of wind turbines (Bennett 2010).

- Precision is required for drilling and finishing of surfaces of the nacelle, the housing that contains the generating components, and the very long blades.
- When putting large turbine blades onto nacelles, robots need precision to ensure correct and consistent alignment. By using intelligent robotics with integrated vision and force control allows the robot to precisely locate and drill holes and finish mating surfaces.
- Robots help overcome the large size of wind turbine assemblies. As Blanchette states (Bennett 2011) "Windmill blades are huge. Processes to the exterior are done less practically with hard automation or by manual labour. Putting a robot on an elevator is much simpler."
- Robots are also used in sanding applications to assure proper surface preparation of the wind turbine blades. This is a critical task because imperfections on turbine blade surfaces cause inefficient eddy air currents over the blade.
- Nacelles and towers are painted, polished, and buffed robotically. With robots, these issues are more consistent and quick, allowing companies to match quality requirements.

In this way, fast and flexible robots allow wind turbine manufacturers to increase production and meet lead times more efficiently, as well as adjust quickly to product changes. The precision and consistent performance possible with robots beat manual methods, vastly improving quality control. Moreover, as wind turbine blades and other components must be finished (painted, sanded, polished, debarred) to perfection, high-quality standards are necessary to maintain the blade balance, extend component longevity, and withstand environmental wear and tear. Robots are helping wind turbine manufacturers achieve the best possible results.

Robotics is also presents in the battery manufacturing which is required to store power produced by wind systems. Wind systems, in the same way that solar systems, are not always generating energy. The use of battery systems to absorb some of this energy as it is generated to be utilized later when the wind is not blowing is a technology in continuous advancement. As batteries contain hazardous elements that people should not directly handle, intelligent robots are required to handle them successfully.

Even though the cost of producing power with wind turbines continues to drop due to the control and optimization techniques improvements and robotics advancements, many engineers feel that the overall design of turbines is still far from optima. Anyway, newer technologies are making the extraction of wind energy much more efficient.

6.4 BioEnergy

Biological energy is the energy that is derived from recently living organisms, such as wood, plant matter, or gases produced by microorganisms. This renewable energy is the so-called bioenergy.

Biological energy sources can be broken down into the categories biomass and biofuel. *Biomass* is produced directly by the living organisms. *Biofuel* is fuel derived from biomass through various processes for the purpose of making transportation of fuel easier, more efficient, or optimized for certain uses.

6.4.1 Biological Energy: Advantages and Disadvantages

Although the burning or conversion of biomass does not fully relieve pollution of the atmosphere, it does have several major benefits. In many regions, biomass is more reliable than solar or wind energy. This is because the energy in plants is captured and stored, while in solar and wind energy, this must be done by manufactured technology.

The main advantages and disadvantages of biological energy are summarized in Borrill (2011). The main advantages are as follows:

- Biomass is a completely renewable resource. Fuel can be produced using grains and plant waste that would otherwise go unused. Many of these plants and grains can be replaced the very next growing season.
- Organic waste exists in abundance and can be used to produce biomass energy. Large amount of solid waste that is currently just dumped into landfills can be used as a source of energy.
- Waste products generated by human activity (paper and household garbage) can be collected and used as biomass to generate energy. This reduces the amount of waste generated and sent to landfills.
- Biomass can be used in many forms: to produce heat, electricity, or other forms of energy. It can be processed and refined to produce alcohols and methane gas, both of which make clean burning sources of energy.
- Biomass energy can save a great deal of money in transportation costs alone. It can be used in the same area in which it is produced. In this way, it is more cost effectively than having huge pipelines or long-distance transmission lines.
- Perhaps, the *most significant advantage of bioenergy* is that it is a potentially renewable natural resource that would help supply energy needs indefinitely.

However, there are some disadvantages to using bioenergy:

- Direct burning of biomass as fuel can release carbon dioxide and other greenhouse gases into the atmosphere (possibly contributing to the problem of global warming). To avoid this effect, converting biomass into a different form of fuel (alcohol or methane) is necessary. This conversion process requires energy input that can make biomass energy more costly than beneficial on a small scale.

- The cost of accumulating and harvesting biomass in its raw form is currently much higher (compared to extracting fossil fuels). It takes time and money to gather and transport biomass to a central point for processing into fuel.
- A biomass power plant would require a great deal of space to accommodate the various stages of collection and conversion of mass into fuel before burning it to produce electricity.
- Water can also be a problem as it would require large quantities to handle the recycling process for waste materials.
- Perhaps, *the major difficulty with bioenergy* is the same problem that has arisen with recycling. People will not demand bioenergy until there is a considerable cost saving in doing so, but there will not be much savings until there is a much larger demand for bioenergy.

6.4.2 Biomass Conversion Process to Useful Energy

There are a number of technological options available to make use of a wide variety of biomass types as a renewable energy source. Conversion technologies may release the energy directly, in the form of heat or electricity, or may convert it to another form, such as liquid biofuel or combustible biogas. While for some classes of biomass resource there may be a number of usage options, for others there may be only one appropriate technology.

Three common conversions to transform biomass to useful energy are the thermal, chemical, and biochemical conversion.

Thermal conversion is the process in which heat is the dominant mechanism to convert the biomass into another chemical form. The basic alternatives of

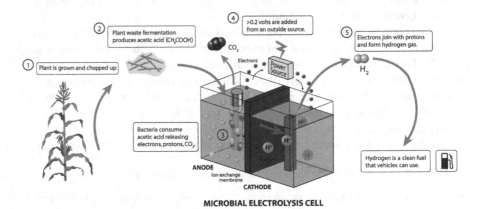

Fig. 6.14 A microbial electrolysis cell can be used to directly make hydrogen gas from plant matter (this is a figure from Wikimedia Commons)

combustion, torrefaction, pyrolysis, and gasification are separated principally by the extent to which the chemical reactions involved are allowed to proceed (mainly controlled by the availability of oxygen and conversion temperature).

There are a number of other less common, more experimental or proprietary thermal processes that may offer benefits such as hydrothermal upgrading and hydroprocessing. Some have been developed for use on high-moisture-content biomass, including aqueous slurries, and allow them to be converted into more convenient forms. Some of the applications of thermal conversion are combined heat and power and cofiring. In a typical biomass power plant, efficiencies range from 20 to 27 % (Dave 2010).

Chemical conversion. A range of chemical processes may be used to convert biomass into other forms, such as to produce a fuel that is more conveniently used, transported, or stored, or to exploit some property of the process itself.

Biochemical conversion. As biomass is a natural material, many highly efficient biochemical processes have developed in nature to break down the molecules of which biomass is composed, and many of these biochemical conversion processes can be harnessed (see the Fig. 6.14).

Biochemical conversion makes use of the enzymes of bacteria and other micro-organisms to break down biomass. In most cases microorganisms are used to perform the conversion process: anaerobic digestion, fermentation, and composting. Other chemical process that converts straight and waste vegetable oils into biodiesel is transesterification (biomassenergycentre.org.uk conversion technologies). Another way of breaking down biomass is by breaking down the carbohydrates and simple sugars to make alcohol. However, this process has not been perfected yet. Scientists are still researching the effects of converting biomass.

6.4.3 Biofuel

One of the major goals of bioenergy is to produce a fuel that has a carbon neutral or even negative impact on the atmosphere. Plants and other organisms take carbon from the atmosphere as they grow. When biomass or biofuels are consumed for energy, carbon is then released back into the atmosphere. This is in stark contrast to fossil fuels which release carbon that has been out of the atmosphere for thousands of years.

Although one of the major goals of biofuels is producing fuel that has a neutral or negative carbon footprint, but this can be offset by certain things. Crops used for biofuel production use carbon from the atmosphere which is released during combustion. This cycle does not take into account nitrogen oxide and other pollutants and green gases released during the production and use of fertilizers, pesticides, and other agricultural aids that may be used.

Land use is another major factor in the environmental impact of biological energy. Growing crops for fuel takes large amounts of land that could otherwise be used for food crops or even preserved in a natural state. Using land to produce crops also depletes soil of nutrients of and can have major impacts on local water supplies due to leaching of chemicals from pesticides and fertilizers.

Biofuels in particular also have an impact on the supply of food crops which can cause changes in the price of food locally and globally. When corn is being used to produce ethanol instead of eaten, the supply of corn is diminished and thus the price increases. Some studies have shown that this impact is not negligible and is already contributing to famine and negative food market fluctuations.

6.4.3.1 Types of Biofuels

All biofuels are derived from organic matter. One exception is cutting edge technology like microbes that convert water and carbon into hydrocarbons (gasoline and more), but these still use organic organisms to produce fuel. Most biofuels are produced from plant crops or waste with sugar cane and corn being popular choices. This biomass is altered by chemical or biological processes to arrive at the final biofuel product.

Probably, the most popular biofuel in the average person's life is direct use of biomass through burning. Wood (biomass) is simply ignited and the heat energy produced can be used for cooking and warmth.

Ethanol is a common biofuel that is most commonly produced by fermenting biomass with specific yeast that consumes sugars and produces ethanol as a waste product. Sugar cane and corn are popular biomass crops used to produce ethanol. A pilot plant to obtain ethanol from biomass is displayed in Fig. 6.15. Ethanol is a

Fig. 6.15 Pilot plant to obtain Ethanol from biomass (this is a photograph from Pat Corkery, Inc/ NREL 16330)

bioalcohol and the most popular bioalcohol, but there are other biofuels like methanol which can also be bioalcohols.

Biodiesel gained a lot of popularity in the last few decades due to the rising price of oil and the fact that it can be used in any diesel engine. Algae biofuel is primarily used in the process of producing biodiesel fuel. Transesterification, the chemical process of making biodiesel, is also a relatively simple and well-understood process. The process is stable and not nearly as hazardous as the production of petro-diesel. The production process also produces little or no noxious gasses to pollute the air around the refinery.

In any algae oil production system, the algae are harvested from the growing process as algae paste. It is then de-watered either by heat drying or de-watering presses. Centrifuges are also another way in which the algae past can be de-watered. The biofuel is then separated from the paste wither by a chemical process or by pressing in a high pressure device such as a screw press. The finished product is algae oil in a form that is then suitable for use in the transesterification process to make biodiesel fuel (see the Fig. 6.16).

The finished product, biodiesel, is an environmentally friendly, renewable fuel with little or no noxious gas release during the process of combustion (see the Fig. 6.17). The production of biodiesel requires one-eighth of the energy required to produce ethanol and is usable in its undiluted state. The demand for biodiesel for use in all sectors now serviced by petro-diesel is projected to grow at an exponential rate.

Biogas is obtained by exposing organic matter to microbes. Methane is one of the most popular biogases. One popular method of producing methane is by using an anaerobic digester (essentially a cylinder with the proper microbes) filled with cow manure allowing cattle ranchers to produce energy from manure.

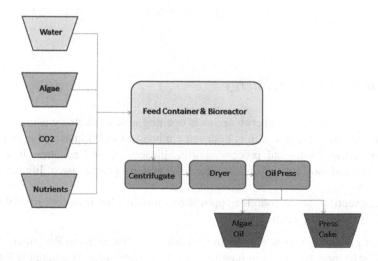

Fig. 6.16 Algae oil production system

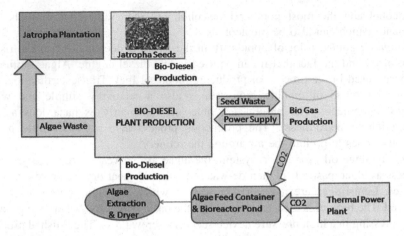

Fig. 6.17 Production of biodiesel

Syngas (synthetic gas) is produced by combustion of biomass in an environment that is low in oxygen. Instead of igniting, when the temperature is increased past the combustion point, the organic matter being used will begin to produce syngas. Normal combustion produces carbon dioxide and water, but syngas is actually carbon monoxide, hydrogen, and hydrocarbons. This mixture is a more efficient energy source when combusted than simply burning the original organic matter and using the heat energy.

There are many emerging biological energy sources that are not being utilized outside of laboratory and research settings. Genetically modified crops and microbes is one of the most promising of these technologies so far. Crops can be genetically modified to produce ideal biomass such as increasing sugar, oil, or carbon levels in the plant. Algae and other microorganisms are being modified to produce biofuels.

6.4.4 Robotics in BioEnergy

There is no doubt that bioenergy sector is opening new markets by using robotics technologies. Investments' aim is to develop advanced techniques and research on bio-engineering, biological process control, biosensing and robotics, bioenergy production, and bio-farming in outer space, by making greater use of life functions themselves.

In this section, some robotics applications looking for investors in biological energy are briefly presented.

- *Virtual production strategies in forestry industry.* (Freund and Rossmann 1999)
  In order to meet the needs of bioenergy in the coming years, sustainable forestry

must become more efficient. Therefore, it has been proposed virtual production strategies, which are well-established in manufacturing, into the forest. By simulating processes beforehand, cost-effectiveness of various strategies can be estimated and the best strategy can be chosen. Furthermore, the visualization allows an intuitive understanding of the simulation results. First steps to integrate robotics know-how with geo-information technology have already been taken by enhancing a forest machine simulator in the way that the virtual forest the machine is working in, may be generated online, based on a topographic map.

- *Floating Robotics Algae Farms.* A new company called BEAR Oceanics (Beargroup) is developing small algal biofuel farms the size of bathtubs that would operate like fleets of robotic, self-contained, biofuel producers, floating in remote-controlled areas of the ocean. That vision aims for self-sustaining robot farms capable of steering clear of boats or ships and relying solely on wind and solar power to grow algae year-round.

The BEAR robotic farms are intended to turn algae sludge into five gallons of biofuel per day, employing a mild electric current to burst the algae cells to release lipid and the thermal depolymerization, using heat and pressure, to turn the algal oil into biodiesel. This will speed up the geological process that created Earth's fossil fuels and all without the risks of drilling for oil or fracking for natural gas. This process allows turning biomass into a biofuel and any chemical is used in the process.

The biofuel farm contained only a small wind turbine, solar panel, and container of green scum packed within a tub-size frame floating on the water. But its design could someday spawn fleets of robotic farms that harness the ocean winds and sunshine to make cheap, algae-based biodiesel fuel for cars, trains, and aircraft.

- *Robotics molecular biology platform for bioenergy applications.* The molecular biological techniques for plasmid-based assembly and cloning of gene open reading frames are essential for elucidating the function of the proteins encoded by the genes (Huhes et al. 2011). High-throughput-integrated robotic molecular biology platforms that have the capacity to rapidly clone and express heterologous gene open reading frames in bacteria and yeast and to screen large numbers of expressed proteins for optimized function are an important technology for improving microbial strains for biofuel production. The process involves the production of full-length complementary DNA libraries as a source of plasmid-based clones to express the desired proteins in active form for determination of their functions. Proteins that were identified by high-throughput screening as having desired characteristics are overexpressed in microbes to enable them to perform functions that will allow more cost-effective and sustainable production of biofuels. Because the plasmid libraries are composed of several thousand unique genes, automation of the process is essential. The design and implementation of an automated integrated programmable robotic workcell capable of producing complementary DNA libraries, colony picking, isolating plasmid DNA, transforming yeast and bacteria, expressing protein, and performing appropriate functional assays is an important task. These operations will allow tailoring microbial strains to use renewable feedstocks for production of biofuels, bioderived chemicals, fertilizers, and other coproducts for profitable and sustainable biorefineries.

References

Abbas F, Abdulsada AM (2010) Simulation of wind-turbine speed control by MATLAB. Int J Comput Electr Eng 2(5):1793–8163

Anderson L, Palkovic R (1994) Cooking with sun-shine (the complete guide to solar cuisine with 150 easy sun-sooked recipes), Marlowe and company, ISBN 15692430

Apte J, Arasteh PE, Huang YJ (2003) Future advanced windows for zero-energy homes. American society of heating, refrigerating and air-conditioning engineers. http://windows.lbl.gov/adv_Sys/ASHRAE%20Final%20Dynamic%20Windows.pdf

Beargroup http://www.beargroup.us/

Bennett BB (2010) A look at robots in alternative energy robotic industries. Available via http://www.robotics.org/content-detail.cfm/Industrial-Robotics-News/A-Look-at-Robots-in-Alternative-Energy/content_id/2433

Bennett BB (2011) Robotics in alternative energy. Available via http://www.robotics.org/content-detail.cfm/Industrial-Robotics-Feature-Article/Robotics-inAlternative-Energy/content_id/3168

Bolton JR (1997) Solar power and fuels. Academic Press, New York

Borrill C (2011) Renewable energy white paper, Southern Hampshire University

Bright D (1977) Passive solar heating simpler for the average owner. Bangor Daily News. http://news.google.com/newspapers?id=beAzAAAAIBAJ&sjid=UDgHAAAAIBAJ&pg=1418, 1115815&dq=improved+by+using+glazing+and+thermal+mass&hl=en

Dave A (2010) Owning and operating costs of waste and biomass power plants. Claverton energy conference

Freund E, Rossmann J (1999) Projective virtual reality: bridging the gap between virtual reality and robotics. IEEE Trans on Roboitcs and Automation 15(3), 411–422

Hammarstrom L, Hammes-Schiffer S (2009) Artificial photosynthesis and solar fuels. Acc Chem Res 42(12):1859–1860

Huhes SR, Butt TR, Bartolett S, Riedmuller SB, Farrelly P (2011) Design and construction of a first-generation high-throughput integrated robotic molecular biology platform for bioenergy applications. J Lab Autom 16(4):292–307. Available via http://www.ncbi.nlm.nih.gov/pubmed/21764025

Jeffrey CS (1999) Row spacing, plant population, and yield relationships University of Arizona. http://ag.arizona.edu/crop/cotton/comments/april1999cc.html

Kaul K, Greer E, Kasperbauer M, Mahl C (2001) Row orientation affects fruit yield in field-grown okra. J Sustainable Agric 17(2/3):169–174

Mark ZJ (2009) Review of solutions to global warming, air pollution, and energy security. Energy Environ Sci 2:148–173

Martin C, Sadrameli SM, Goswami DY (2005) Comparison of optimum operating conditions for a combined power and cooling thermodynamic cycle. In: Proceedings of the ISES solar world congress, August, Orlando, FL

Muhs J, Oak Redge National Lab. (2000) Design and analysis of hybrid solar lighting and full-spectrum solar energy systems. SOLAR2000 conference. http://web.archive.org/web/20070926033214/http://www.ornl.gov/sci/solar/pdfs/Muhs_ASME_Paper.pdf

National Instruments (2008) Wind turbine control methods. http://zone.ni.com/devzone/cda/tut/p/id/8189

Neammanee B, Sirisumrannukul S, Chatratana S (2007) Development of a wind turbine simulator for wind generator testing. Int Energ J 8:21–28

OECD/IEA (2011) Solar energy perspectives: executive summary. Available via http://www.webcitation.org/63fIHKr1S

Pasqualetti MJ, Miller BA (1984) Land requirements for the solar and coal options. Geographical J 150(2):192–212

Philibert C (2005) The present and future use of solar thermal energy as a primary source of energy IEA. Available via http://www.webcitation.org/63rZo6Rn2

Somerville R (2007) Historical overview of climate change science. Intergovernmental panel on climate change. Available via http://www.ipcc.ch/pdf/assessment-report/ar4/wg1/ar4-wg1-chapter1.pdf

University of Michigan. Perspectives on alternative energy. Available via http://sitemaker.umich.edu/section9group4/wind_energy

U.S. Department of Energy (2011) Advantages and challenges of wind energy. Available via http://www1.eere.energy.gov/wind/wind_ad.html

Vermass W (2007) An introduction to photosynthesis and its applications. Arizona State University. http://photoscience.la.asu.edu/photosyn/education/photointro.html

Wasielewski MR (1992) Photoinduced electron transfer in supramolecular systems for artificial photosynthesis. Chem Rev 92:435–461

Weiss W, Bergmann I, Faninger G (2005) Solar heat worldwide: markets and contributions to the energy supply. International energy agency. http://www.iea-shc.org/publications/statistics/IEA-SHC_Solar_Heat_Worldwide-2007.pdf

Chapter 7
The Future of Smart Domestic Environments: The Triad of Robotics, Medicine and Biotechnology

José Antonio Díaz and M. Rosario Hilde Sánchez Morales

7.1 Introduction

"New research directions in robotics technologies promise wide-scale adoption of robots in all aspects of life—from industrial manufacturing to use in professional and domestic service environments. However, applications for small company or personal use are still at the research stage. Though modern science fiction has embedded in our psyche the idea of automated machines with more or less human characteristics, the robotic equivalent of the personal computer, as in the domestic service robot for personal and family use in a household, remains a long-term goal" (Forge and Blackman 2010).

The aim of this paper is to analyse the impact of long-term technological innovation in the home environment. We analyse the transformation that households may experience due to the increased use of technology. With respect to long-term technology forecasting and social applications, we are discussing conjecture and an acceptable level of probability; we consider speculative estimates of future technologies and the potential uses of these technologies by citizens.

Often, technology experts, particularly those with technical training, say that if technology can do something, it will. There is some "technological determinism" in this approach, and, indeed, the history of technological innovation shows an apparently endless number of business failures involving products that were not accepted by users. In speaking of the future smart home, we suggest just the opposite view: just because something is technologically possible does not mean that it will be achieved. Other variables, such as those related to values,

J. A. Díaz (✉) · M. R. H. Sánchez Morales
Department of Social Trends of the National Distance Education University (UNED),
Universidad Nacional de Educación a Distancia, Madrid, Spain
e-mail: jdiaz@poli.uned.es

M. R. H. Sánchez Morales
e-mail: msanchez@poli.uned.es

A. López Peláez (ed.), *The Robotics Divide*,
DOI: 10.1007/978-1-4471-5358-0_7, © Springer-Verlag London 2014

uncertainties, the real needs of people, patterns of behaviour and personality, are involved: of these, culture is among the most important.

The home is a socially defined space; if technology is difficult to anticipate, it is even more difficult to understand the wishes and needs of consumers and users within the next 20 years.

For decades, socio-technical scenarios have been constructed to predict technological breakthroughs and the solution to important problems of everyday life. A wide social awareness of the benefits of technological revolution exists, linking technology with progress, prosperity, wealth and comfort. In the opinion of experts in the study of the future, the end result of this process is the creation of the electronic home and even of the smart home, which is an interesting and disturbing concept.

This concept of a "smart home" can be used in a broad sense to describe tools that help in the home or, in a deeper sense, tools that lead to a radical transformation of the home, thereby redefining a conceptual and multifaceted social space that is able to assist residents in various activities and anticipate their wants and needs (Díaz 2010).

One of these futurists, Alvin Toffler, who anticipated the arrival of the "Electronic Cottage" by four decades, thought that the home would be the place for education, health care, leisure and entertainment and even work; i.e., new production systems might shift literally millions of jobs from factories and offices to their original location, the home (Toffler 1984).

Toffler, in his book "The Third Wave", foresaw the collision between organisations typical of the 2nd and 3rd waves. The InfoSphere of the age of information creates a new mode of production: turning the home into the new electronic environment and setting the home as the centre of society (Toffler 1984:231).

Toffler was well aware of the scepticism met by his forecast, which raised a number of objections, the last of which read literally, "What, stay all day at home with my wife [or husband]?" (Toffler 1984:232).

However, in his opinion, there was also strong resistance to the reverse transition, from the home to the factory. The family had been the locus of production for thousands of years. There were social conditions and a new mode of production, transforming life and work, building the large urban and industrial centres and the characteristic organisation of industrial society.

Further research and analytical perspectives on the future highlight that housing will be particularly sensitive to the impacts of technological change and that the transformation of production systems, consumption and culture will be reflected prominently in the houses of the future (Tezanos and Bordas 2000). Toffler foresaw an inevitable shift towards globalisation, the networked society, individualism and the electronic home in modern societies (Garson 2006). Now, more than 30 years later, the question is as follows: how will the smart home of the future develop?

Modern home computing began to spread in the 1980s, initially focused on video games and word processing, but later developing into educational tools and work. However, one could not talk in terms of a significant transformation of the social model and the marriage of computing and communications until the advent of the Internet in the 1990s (Castilla and Díaz 2008).

7.2 The Background of Technological Change

Current technological forecasts indicate a turning point from 2015, when a significant increase will occur in sectors related to services for people in households. As noted by Forge and Blackman, "Robots for elderly care, as a domestic service category, will take longer to take off for reasons of cost/functionality and acceptance by ordinary people (especially outside Japan). Such robots will only become market priced for major growth well after 2015. If they can provide good functionality and provide value for money, simple care robots could grow rapidly to be a mainstream product category by 2020, especially if the technology for receiving spoken instructions can be developed. In contrast, more complex domestic service robots for ordinary people who are not elderly, frail or disabled may only grow after 2020 when price and performance give real advantage over human activities for domestic chores. For most households, even in OECD countries, this is likely to remain a restricted market for the foreseeable future, not becoming mainstream until later in the century" (Forge and Blackman 2010:48) (see Fig. 7.1).

Indeed, today, we are at the pre-launch stage of the new technological paradigm. As noted by Carlota Perez, the origin of this shift is a technological revolution; a revolution resulting from the integration of two major changes: one, the

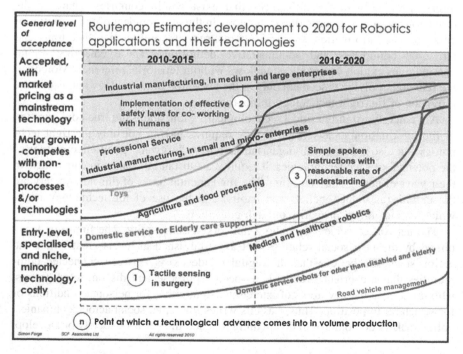

Fig. 7.1 A roadmap of progress and market entry by segment. *Source* Forge and Blackman (2010), o.c., p. 48

information revolution, and two, the organisational revolution. It is difficult to understand this process of creative destruction (Pérez 2000).

Many problems involving the social applications of new technologies arise from the necessary reconfiguration of social systems. The new techno-economic paradigm is a change in the way of doing things, and the new type of organisation requires a new logic. We are on the threshold of a new economy (Tapscott 1995; Tapscott and Williams 2007).

Carlota Perez points to the fact that the jump is not based on computerisation and Internet use, although these certainly play roles in the changes in productivity and quality. However, the old form of organisation does not permit taking advantage of the new technologies. New principles, practices and organisational models (Pérez 2000:5) are required.

Another problem is the digital divide, the inequality that occurs because of the conditions of the use and disposal of information technology in different social sectors (Servon 2002). The term digital divide is the opposite of digital inclusion, that is, the policy that aims at homogeneous development and inclusivity regarding the benefits of information technology.[1] In fact, we should also consider the various aspects and areas of impact of such technologies; we could discuss the digital divide between countries and the gap that occurs within a single country, affecting society (Norris 2001). We believe that the most appropriate term to use when discussing the information society is "the information community", which refers specifically to the differences that exist within countries regarding the provision and use of information technology in the development of the activities of citizens. The information society is composed of communities that transcend territorial boundaries and administrative definitions. These emerging digital communities exhibit communication flows without territorial reference. Moreover, the real digital divide is seen not only in the provision of technology but also in the provision of the right technology infrastructure.

Are emerging rightly critical technological vision, about the true configuration of the information society. This society requires having not only the right technology but also a new social organisation, a new "orgware", that effectively uses the potential of such technologies. It requires a substantial change in companies, in their ways of acting and managing. The fundamental bases of this new society are the culture (as values, beliefs and habits) and the use of the technology itself, which is transmitted primarily through education.

The networked society is really a networked community. The future is determined by the new social relations that are established at levels greater than the nation state. We believe that the digital divide exists between connected communities that are oriented towards a specific purpose and disconnected groups, who do not use the new tools of interconnection, either for personal conditions or for structural imposition. Thus, Castells writes, "People, social actors, companies, policy makers do not have to do anything to reach the network society or Develop.

---

[1] http://cdieurope.eu/about/cdi-global/

We are in the network society, although not everything or everybody is included in its networks. Therefore, from a policy standpoint, the key question is how to proceed to maximize the chances for fulfilling the collective and individual social projects that express needs and values under the new structural conditions" (Castells and Cardoso 2005).

Social development depends not only on the existence of a particular technological infrastructure but also on a set of integrated actions that foster a qualitatively different kind of society. In this sense, the institutional efforts that have been made to promote the knowledge society do not have the desired result. Therefore, while considering the European Union's efforts to promote competitiveness, Manuel Castells notes, "The European Technological Infrastructure considerably improved, but effects on productivity, on learning, on creativity, and on entrepreneurialism, were very limited. Because this is acting on the developmental potential specific to the network society requires a combination of initiatives in technology, business, education, culture, space restructuring, infrastructure development, organizational change, and institutional reform. It is the Synergy between these processes that acts as a lever of change on the mechanisms of the network society" (Castells and Cardoso 2005:17).

7.3 The Domestic and Personal Environment

The home has gone from being a "closed" space that is reserved for family relationships and personal social relationships to being an "open" place, that is, a channel that is capable of erasing the borders of the internal and external. The home has become a potentially public space that includes possibilities for any social and economic activity. We can say that the "Evolution of households that took place during the decades of the fifties, sixties and seventies was largely a revolution in design, furniture and brick, while the revolution of this decade will have a greater component of technology and equipment" (Tezanos and Bordas 2000:19–20).

The terminal that will allow the indicated processing is the computer or another equivalent electronic device that has multiple functionalities. The most important transformation applies to the content of the new telematic channels, which all occur at a distance: training, working, buying, diagnosis and leisure.

Being present at a distance allows work to be performed from home, and various forms of teleworking are possible: being in the office on certain days of the week or working for different companies from home.

Moreover, no great confidence exists that the creation of smart environments is profitable (Abadie et al. 2008).

The concept of the smart home environment implies the convergence of two related types of development: smart homes and ubiquitous computing (Tezanos and Bordas 2000:277). This type of space can be defined as a house equipped with computers and information technology, which anticipates and responds to the needs of the residents and attempts to promote comfort, convenience, security and entertainment, through the management of technology within the home and connected to the global environment (Tezanos and Bordas 2000:277).

In recent years, the scientific literature relating to homes has discussed the home and intelligent building. The word "smart" has been supplemented by new concepts, such as the "aware" home, integrated environments, or live, interactive environments. Similarly, there has been a significant change away from technocentrism and towards paying more attention to the person and a flexible relationship with technology, considering the great potential of technology to consider the real needs of users.

Technological innovation is important, but some factors, such as social changes, are likely to influence the uptake and increased use of robots; for instance, "the pressures of an ageing population bring new demands at a mass-market level, perhaps changing the future direction of the whole robotics market away from its industrial manufacturing roots towards new domestic service applications. Today, this is inhibited by price. Researchers at the Fraunhofer Institute, for instance, believe that a personal domestic robot with more than basic functionality would need to be available to consumers at a price similar to a good quality car (approximately €40–50 k) to enable significant take up. However, a price point of below €30 k (except for early adopters, perhaps €50 k) is possibly the limit that can be afforded for widespread take-up. Other potentially significant robotics technologies such as the exoskeletons for the mobility of the aged, now in early forms such as the Japanese HAL exoskeleton robot, have a market price somewhat higher than this, in excess of €50 k. HAL only rents currently, and it is likely that new business models will be necessary" (Forge and Blackman 2010:44).

In the social context that defines the emergence of the smart home, there are many unknowns and variables, doubts and uncertainty; however, the main trends driving the application of ambient intelligence technologies in housing can be identified as follows (Abadie et al. 2008:277):

- acceleration of rhythm of everyday life, hectic and busy lifestyles, growing demand for efficiency and flexibility in daily routines,
- breaking up of the boundaries of time and space (increasing telepresence),
- ageing of population, leading to a demand of elderly living longer in their homes,
- increasing demand of security and safety (e.g. due to rising crime rates),
- growing pressures to curb environmental problems and to save energy (to promote sustainable development and to compensate high prices of energy),
- increasing search for experiences and meanings (leading to building homes as media/entertainment centres),
- increasing need for home as a sanctuary (home as dedicated to privacy, rest and relaxation),
- "technological" way of life, i.e., lifestyles that take technologies for granted.

People have ambivalent feelings towards technological applications in the home. First, the smart home must provide comfort. However, concern remains regarding the dehumanisation that result from technological applications and the lack of confidence in and control over technological solutions (Abadie et al. 2008:278).

When we discuss the smart home, we are referring to the process of introducing new technologies with different configurations (Aldrich 2003):

1. Households that contain intelligent objects: these are households that have individual technological applications that work intelligently.
2. Households that contain intelligent objects and communication: these households contain devices that work intelligently and communicate with other devices to exchange information and increase functionality.
3. Adaptive households: These are households in which activity patterns are recorded and the resulting data are used to anticipate the needs of users and control technology (Mozer 2005).
4. Active households: In these households, the activities of the people and household objects are monitored constantly, and the information is used to control the technology and provide for human needs (Abowd et al. 2002).

This technology is currently feasible. The question is what the average user really needs. According to experts, the problems are determining how many domestic functions are valuable and the fact that many users will have the technology and forget to use it (Tezanos and Bordas 2000:53).

Home automation applications have a long history; they initially focused on the development of services and products related to the control of simple and integrated electronic applications (e.g. fire alarms). Currently, the majority of home automation systems are not "intelligent". Home automation technology has not yet attracted sufficient interest for its use (Abadie et al. 2008:280).

In the time frame under consideration (20 or 30 years), innovations such as DNA computing and nanomolecular computers, which will substantially increase the ability to capture, store, process and disseminate data (information), could occur. Grid-based supercomputing and parallel architecture will be a widespread reality in the shorter term: first, in large industries and research centres, and then in the domestic environment.

There is another remarkable phenomenon, which we would call *anthropomatic*, which refers to the implantation of technology in humans (cyborgs). Far from the futuristic visions provided by Robocop, society currently assumes that electronic implants are possible and encourage their improvement and human development (The Academy of Medical Sciences 2012).

Devices are currently implanted for therapeutic use (for example, pacemakers, robotic arms and hearing or visual implants); however, such devices are potentially applicable to improving human capabilities in general. This will result in two important issues: defining the ethical and moral boundaries of such innovations, and in the longer term, defining the freedom of the person and the control of human behaviour by machines.

In the same way, "researchers worldwide have been pursuing the goal of humanoid robots, potentially useful as domestic household servants, carers and helpers" (Forge and Blackman 2010:9).

A future challenge lies in the human–robot interaction (HRI). As indicated by the experts, "a key attribute of robotics is the ability to communicate and share goals and information with humans, so robots can meaningfully become part of the human environment. HRI is based on studying the communications processes between humans and robots. It brings together approaches from human factors, cognitive psychology, human–computer interactions, user interface design, ergonomic and interaction design, education, etc., in order that robots can gain more natural, friendly and useful interactions with humans. (...) HRI for close co-working with humans include: home support for care of the elderly, rehabilitation of the frail, hospital care support, education, and emergency first-responder support" (Forge and Blackman 2010:19–20).

Major technological innovations in the home in the coming decades

Development of the Internet is a key element in the evolution of the information society (Castilla and Díaz 2008). Future Internet transformations define the development of new social applications, especially in the education system and the relationship between citizens based in the home and the health system, public administration and businesses.

In this sense, expert technologists believe that the main innovations of these technologies in the next 10 years will seek the integration of previously separate systems into a single multipurpose device. Additionally, human–machine interactions will improve upon using natural language, and better storage and transmission services will be developed with increased security and mobility (Table 7.1).

These developments and, above all, their application will depend largely on the circumstances or environmental constraints; however, these innovations will affect the home in one way or another and facilitate the implementation of many activities.

Table 7.1 Eight outstanding innovations in science, and information and communication technology

	Value
1. The unification of TV/Internet/radio/telephone systems	25
2. Reliable recognition of the human voice such that instructions can be given to machines, text can be typed, etc.	21
3. Image, voice and data processing systems with higher performance	21
4. Integration with household technologies	21
5. Efficient and secure communications encryption	18
6. Expanding Google-type search engines, making them more selective, fast and accessible	17
7. Development of wireless systems that facilitate the ubiquity of connectivity	16
8. Improved system information storage and transmission at very high speed	14

Source Tezanos JF et al. (2005)

Structural conditions can facilitate the future application of these technologies at home, such as investment promotion and institutional policies; however, we consider that one of the key elements in this process is culture. Culture can delay change by maintaining traditional patterns of behaviour. In this sense, the "digital generation" will be crucial to the transformation to a knowledge-based society (Díaz 2009). Indeed, the main limitation to the development of science and technology is the lack of technological culture among the population. However, experts believe that "Effort should be focused on the largest unexploited opportunities, e.g., in food processing, high-tech industries and professional and domestic services" (Forge and Blackman 2010:12).

Experts say, "These are robots that operate semi- or fully autonomously to perform services useful to the well-being of humans, for instance: Robot butler/companion/assistants; Vacuuming, floor cleaning; Lawn mowing; Pool cleaning; Window cleaning; Robotised wheelchairs; Personal rehabilitation and other assistance functions. The Husqvarna Automower was the world's first robotic lawn mower, with over 100,000 units sold since 1995. iRobot's Roomba is an autonomous robotic vacuum cleaner that is able to navigate a living space and its obstacles while vacuuming the floor. The Roomba was introduced in 2002 and has sold over 2.5 million units" (Forge and Blackman 2010:23).

7.4 Robotic, Medicine and Biotechnology Applications

The applications of microrobotics and macrorobotics in medicine can be grouped into tree areas of priority (Najarian et al. 2011).

7.4.1 Robotic Surgery or Telesurgery

The purpose of robotic surgery is the improvement and restoration of patient health. Robots can be managed remotely by experts, giving rise to robotic surgery or telesurgery (Marescaux et al. 2001) and computer-assisted surgery (Lum et al. 2009). This field originated in July 1992 with the Defense Advanced Research Agency[2] and Advanced Biomedical Technology,[3] designed to treat the wounds of soldiers on the battlefield. This technology is based on two fundamental concepts, virtual reality and cybernetics, and is seen as twenty-first century surgery. It should be clarified that while robotic surgery is the robot, once programmed, performed the operation itself, there are robots in telesurgery procedures performed entirely

---

[2] http://www.darpa.mil/
[3] http://www.advbiomed.com/

under the supervision of the surgeon. The future should see the development of fully autonomous robots, although these are currently in the experimental stage.

In this context, endoscopic surgery has gained a high profile because it is used to address issues in various parts of the body (e.g. arteries, veins, organs and bones) (Hosseini et al. 2006). Endoscopic surgery has evolved towards minimally invasive techniques (Westebring van der Putten et al. 2008). The first robotic arm that was designed for laparoscopic surgery in the USA was built in 1993; currently, more advanced versions exist, some of which operate by voice command with memory positions. The first of these include the Da Vinci robot (Ballantyne and Moll 2003) and Zeuss; these follow the CASPAR,[4] the Inch-worm,[5] the Probot, Robodoc, Minerva (Villorte et al. 1992) and AESOP, as well as coaches for various types of surgical interventions.

Robotics is just one of the technologies that can contribute to the development of surgical techniques, due to their high precision, the stability of their movements, their ability to be programmed and managed by remote control and their high level of asepsis. Robotic surgery uses robots, not to develop therapeutic procedures but as auxiliary systems for execution. Robotic systems are, therefore, automated systems that are complementary to the surgeon's performance. Therefore, we can say that in the medium and long term, advances in communication technologies and information, robotics, and the possibility to visualise objects in three dimensions inside the human body (using TACs, MRI and ultrasound), have transformed the practice of medicine, thereby introducing new possibilities regarding diagnosis, therapy and high-precision surgical techniques. For now, the technology is restricted by its high economic cost (e.g. a Da Vinci machine costs between $ 1,200,000 and $ 1,500,000) and the need to properly train health professionals in its use.

Most of the surgeries performed by robots involve the following specialties: vascular surgery, gynaecological laparoscopic surgery (e.g. hysterectomy, gynaecologic cancers and removal of the lymph), endoscopic cardiac surgery, laparoscopic surgery, neurological, intracranial and ophthalmic surgery (e.g. damage to the retina), jaw surgery (while implanting hearing aids), spinal surgery and orthopaedic surgery (e.g. hip replacements). Robotic equipment has also been developed that assists surgeons in delicate, microscopic operations (e.g. in ophthalmology, neurosurgery, otolaryngology, cardiology, obstetrics, gynaecology, urology and gastroenterology). Finally, we note the increasing use of microrobots for medical purposes, such as capsule endoscopy and colonoscopies.

In November 2004, a company introduced a capsule robot capable of exploring the human body. This is a smart device, currently used by the most advanced medical groups in the world, which is introduced into and can move through the body while controlled from the outside by specialists.

There are many advantages of using robots in surgery (see Table 7.2).

---

[4] http://www.goetgheluck.com/PDF/medical_robotic.pdf

[5] http://groups.csail.mit.edu/drl/wiki/images/c/cd/fulltext.pdf

Table 7.2 The benefits of using robots in surgery

1. Increases the motor skills of the surgeon and allows easy handling
2. Reduces the time required for operations
3. Extends the surgeon's professional life
4. Operates with greater precision of movement, security and flexibility
5. Simplifies the use of invasive surgical techniques
6. Allows the possibility of programming similar operations using only minor adjustments
7. Makes it easy to follow precise orders and is secure
8. Allows remote operations
9. Allows shorter stays and reduces postoperative problems
10. Allows more rapid patient recovery

7.4.2 Prosthetics and Orthotics

The aim of these is to replace human members and drop functionality equivalent prosthesis. Its development requires, in most cases, the use of robots that imitate natural movements.

It is a scope, which are happening continual advances in recent years, combined with the use of microelectronic signals in the hands, arms, legs, etc., and functional electrical stimulation (in the case of exoskeletons for tetraplegics). However, brain–machine interfaces are already in the experimental phase; these interfaces are able to help paralysed patients to move robotic prosthetic arms or legs. These interfaces are based on the idea that thoughts and movements are electrical phenomena. A time may come when computers are able to predict the movement of the legs, arms or body based on brain patterns. It may be possible in the future to activate robotic prostheses with the mind.

7.4.3 Care Robots

7.4.3.1 Robots for Disabled People and the Elderly

Robots and tools can be developed based on an improved understanding of human psychology to create friendly robots that act "as humans", equipped with basic features for entertainment and assistance.

In this context, the O-Bot Care system is very relevant.[6] This system has the following characteristics:

- Responds to the voice
- Enables communication with welfare services (social and health)
- Approaching the food and drink

---

[6] http://www.care-o-bot.de/english/

- Allows the control of temperature, light and alarms in the house
- Provides personal assistance with dressing, walking, lifting, etc.
- Enables the monitoring of vital signs of people, sending medical and fire alarms in cases of emergency

The ROBOR LX2[7] was sold in 2005 to help seniors who had just undergone knee replacement and patients who were recovering from heart attacks or strokes. This system consists of a mechanical arm that assists patients in moving their legs, thereby allowing them some mobility. Another example is the robotic assistant for the elderly (Pearl),[8] which has a human face and uses software that lets you move and issue verbal reminders regarding tasks such as taking medicine and attending hospital dates. Finally, appropriate rehabilitation robots (termed "tele-rehabilitation") use virtual systems that connect patients with clinics through PCs, webcams and microphones (Krebs et al. 2007).

7.4.3.2 Hospital and Research Robots

The purpose of these robots is to eliminate the monotonous and repetitive tasks that are performed in hospitals. For example, robots are used for radiation treatments; these robots plan the locations and sequences of radiation used with remarkable precision and are highly mobile. HelpMate robots serve food to patients and transport medicines and samples for analysis. Robots assistants, whose main function is to enable doctors in their interaction with patients, recently performed operations without being present. In September 2004, the first robot of its kind was installed at Davis Medical Centre at the University of California; the body of this "RoboDoc" consists of a camera, a television screen and a microphone. The camera can be zoomed to view vital signs and surgical incisions. The machine operates through a hospital's wireless network and only requires a computer and a camcorder for use. This robot can be handled even from outside the hospital.

Finally, the growing presence of robots as research assistants should be mentioned; these robots are used in the discovery of new drugs.

7.4.3.3 Nanotechnology

One of the most important areas of medicine is nanotechnology. In 1959, the Nobel Prize winner Richard Feynman stated, "the principles of physics do not deny the possibility of manipulating atom by atom. The problems of chemistry and biology could be avoided if we develop our ability to see what we are doing and to

---

[7] http://www.yaskawa.co.jp/en/company/csr2008/06.htm

[8] http://www.ri.cmu.edu/research_project_detail.html?project_id=347&menu_id=261

Table 7.3 Some of the main and future applications of nanomedicine

Molecular diagnostics (biochips/biosensors):
- For the diagnosis of genetic diseases.
- For the detection of viruses.
- For the detection of cancer.

Delivery agents for "smart" medicines:
- For the treatment of diabetes.
- For the treatment of cancer.
- For the treatment of viruses and infectious diseases.
- For the treatment of kidney diseases.

In neurological therapeutic processes:
- To restore senses such as sight and hearing.
- To slow the progression of Parkinson's disease.

In the therapeutic response to trauma:
- To heal osteoporosis and broken bones.

---

make things level atom". This was the first time that someone had mentioned the possibility of intervening on the atomic scale in public.

Nanotechnology holds the promise to diagnose diseases and treat them at the cellular or molecular level from within the body. In short, nanotechnology envisions a therapeutic setting involving the use of small machines to stop disease (Drexler et al. 1991).

The application of nanotechnology to medicine is one of the most promising among the foreseeable technological advances in medicine and will lead to a new scientific era. Certain fields may experience a revolution, especially those related to monitoring, imaging, tissue repair, control of the evolution of disease, protecting and improving human biological systems, pain relief, drug delivery to cells and, in particular, to combat and treat cardiovascular disease, diabetes, cancer and Alzheimer's and Parkinson's diseases. Specific applications and future can be identified as follows (Table 7.3).

7.5 Temporal Scenarios

Around 2025, the maximum level of innovation of these technologies will be achieved in areas such as public administration and business. This will make major transformations likely in the form of relationships within society and promoting the intensification of the Information Society. However, we will have to wait another two decades, until the 2050s, to experience widespread changes in the field of privacy, when a different technological culture will emerge (Table 7.4).

Table 7.4 Technological trends related to the smart home in Spain

Trends	2015	2025	2050
● Percentage of the population connected to the Internet (%)	60	90	100
● Percentage of people who usually work from home (%)	10	20	38
● Percentage of households that have integrated electronic household services (%)	5	20	50
● Percentage of households that are connected to the Public Health System (%)	3	20	30
● Percentage of people who will interact with government through the Internet (%)	50	70	90
● Percentage of people who obtain medical advice from home through the Internet (%)	5	20	30
● Percentage of users who are connected to leisure activities for more than three hours each day (%)	20	30	40
● Normal laptop battery life normal laptops in hours (h)	5	15	30
● Percentage of phones that come equipped with UMTS or higher (%)	50	100	100
● Percentage of mobile phones equipped with cameras and image transmitters (%)	50	90	100
● Percentage of people who use a PDA (%)	30	40	50
● Percentage of TVs with interactive access to the Internet (%)	10	60	80
● Percentage of books that are published electronically (%)	10	20	35
● Percentage of university content that is accessible via the Internet (%)	45	55	80

Source Tezanos et al. (2005) o.c., p. 86

Technological trends in future decades:

1. A sustained increase in the number of Internet users. In the 2020s, the number of Internet users will reach a significant percentage. The generalisation of this medium will allow for the development of social applications for the home.
2. Improved ICT equipment, in terms of capacity, range, portability and cost. These technological improvements will allow a substantial increase in the number of teleworkers, i.e., people who spend most of their working hours in the home. The predictions are that this workforce percentage will reach 20 % in the 2020s and just under 40 % in the 2050s.
3. The most important applications are related to health, education, road safety and the dissemination of information and knowledge. In fact, it is expected that approximately 20 % of Spanish households will have Electronic Integrated Systems at Home in 2025 and that 50 % will do so in 2050.
   Regarding health, the forecast for 2025 is that 20 % of households will be connected to the public health system, and this figure will reach 30 % in the 2050s. Consultations from home will also experience a significant increase, reaching 30 % in 2050.
4. The areas that will drive these developments will be education (schools and universities), business and public administration (resulting from the strong development of e-government). The institutions mentioned are in fact the drivers of the information society.

5. In the 2nd and 3rd decades of this century, Spanish households will experience the biggest changes in ICT. In addition to the above-mentioned processes, other areas of household activity will also be transformed, such as "leisure and entertainment"; 40% of the population will spend more than 3 h a day connected to the Internet for this purpose in 2050.

In short, the smart home is a reality in modern societies. A significant portion of current technological innovations has a significant impact on shaping the personal and domestic environment. The new values of society, such as concern for the environment and sustainability, the importance of security, information workers, or the growing importance of entertainment and leisure in the lives of people, are elements that guide the innovation process in households.

The data analysed above indicate that during the next two decades, we will approach the knowledge society and, therefore, the redesign of the human habitat. The home of the future cannot escape the transformation that will occur in society.

The problem of the digital divide will likely remain in the near future. The state has an important role to play in the development of an inclusive social model that is based on a new way of organising technology, as Castells notes (Castells and Cardoso 2005). However, we believe that the emerging society is a social network that will overcome the dynamics of the state itself. Significant change will come from within the interstices of the social system, and then, the state will have to adapt; indeed, in certain circumstances, the state has a role to play in retarding the changes that lie ahead. In fact, the forecast is not good. The divide between developing and developed countries and connected and disconnected communities could get deepened. As Himanen noted, "If we carry on with" business as usual, "inequality and marginalisation will continue to become aggravated Both Nationally and globally" (Himanen 2005).

The European model of transition to the knowledge society imposes some limitations on the prevention of a digital divide. As Himanen says, "the combination of the current European information society and the Welfare State, has the danger of the dead hand of passivity. According to this scenario, all the people keep protecting industrial era structures of the welfare state, but they do not recognise that the future of the welfare state is possible only if the welfare state is reformed with the same kind of innovativeness that the information economy has gone through" (2005:344).

The condition for the development of an inclusive knowledge society is education. As Himanen noted, "The success of the information society and the provision of equal opportunities in the Welfare Society are, eventually, based on an inclusive and high-quality education and training system. In the information society, where learning continues throughout our lives, schools should not only distribute information but also, and equally Importantly, build self-confidence and social skills, as well as help pupils to fulfil Identifying themselves by their talents and creative passions. In addition, the challenge of lifelong learning in the information society must require that people learn to learn-become able to Identify problems, generate ideas, apply source criticism, solve problems and work

Table 7.5 Forecasts of Japanese and Spanish experts in robotics applied to medicine and nanotechnology (1)

2011–2015	2016–2020	2021–2025	2026–2030	2050 and more
- Electronic devices will exist that stimulate the brain's pleasure centre, having similar effects to psychoactive substances	- Hospitals will often implement robotic prostheses that replace the lack of human limbs	- Frequent use will be made of neural chips with gigabit storage capacity	- Most of the hospitals in Spain will use nanotechnological devices to diagnose and cure diseases	- Computers will be able to "read" the electric and magnetic information recorded in the human brain
		- Artificial muscles will be developed - Development of artificial organs composed of biological cells and artificial objects		
- Development of medical diagnosis and treatment with micromachines that are able to travel within the cavities of organs	- Frequent use of devices implanted under the skin for medical diagnostics	- Design of home robots that can learn the habits of their owners		

Source: National Institute of Science and Technology Policy 2001 and Grupo de Estudio sobre Tendencias Sociales (GETS) 2002, 2005, 2011

Table 7.6 Forecasts of Japanese and Spanish experts in robotics applied to medicine and nanotechnology (2)

2011–2015	2016–2020
– Development of micromachines that make gastrointestinal examinations	– Frequent use of microrobots for internal examinations of the human body using integrated sensors
– Development of standard noninvasive techniques to obtain blood sugar levels using intracutaneous sensors	– Development of minimally invasive techniques using microrobot machines in most medical operations
– Frequent use of automatic devices to perform diagnostic radiology	– The development of remotely controlled microdevices for treatment of the thrombus
– Development of effective radio sensors for the treatment of cancer	– Routine use of sensors capable of stimulating human nerves without being implanted into the skin
– Routine use of artificial implants as hearing aids	– Operating through a remotely controlled system of micromachines equipped with sensors
	– Development of electrical circuits similar to self-organised neural networks
	– Development of implantable artificial kidneys
	– Routine use of artificial pancreases
	– Routine use of artificial livers
	– Progress in bioinformatics that integrates and enables the widespread use of data related to the life sciences and the development of networked virtual laboratories
	– Development of artificial eyes that can be connected to the nerves and brain cells

Source Idem

together with Other people. Teacher training should pay more attention to these matters" (2005:358).

As indicated, the process of implementing processing technology and society occurs at a high speed. Thus, highly competitive networks reach across the borders of countries; other, mid-level networks do not reach that level, and a significant number of people remain outside the network. It is not easy to define the dimensions of the network society because the network society does not match the dimensions of the states; rather, a relationship exists between the networks that comprise a network system. The state must transform its traditional role (Castells and Cardoso 2005:19). The state must be a provider of infrastructure for the self-organisation of civil society. As Mulgan says, "It is rather that states are reshaping themselves to be less structures that provide services directly or achieve outcomes; instead they are becoming more like infrastructures, orchestrating complex systems with greater capacities for self-organization, and engaged in co-creation of outcomes with citizens and civil society" (Mulgan 2005).

People may consciously resist the changes that will occur, but most will accept them. In that sense, and regarding the situation in our country, there has been a change in technological culture. As mentioned earlier, modern generations have a culture that favours the introduction of technology into their lives and perceive technology as an opportunity or an advantage rather than as a threat to our humanity. This is a critical factor, and, in our opinion, it is a necessary condition to realise the potential of the scientific and technological revolution that characterises most advanced countries.

Moreover, advances in medical biotechnology have placed us at a new stage. The medicine involved will be eminently predictive/preventive, and robotics and nanotechnology will play a major role (see Tables 7.5, 7.6) (Rogozea et al. 2010).

References

Abadie F, Maghiros I, Pascu C (eds) (2008) European Perspectives on the Information Society: Annual Monitoring Synthesis and Emerging Trend Updates. ICT tools and services in intelligent domestic and personal environments, European Commission, p 275. https://observatorio.iti.upv.es/media/managed_files/2008/12/10/JRC47923.pdf

Abowd GD, Bobick I, Essa E et al (2002) The aware home: developing technologies for successful aging. In: Haigh K (ed) Automation as caregiver: the role of intelligent technology in elder care. Papers from the AAAI Workshop, Menlo Park: Association for the Advancement of Artificial Intelligence (Technical Report, WS-02-02), pp 1–7

Aldrich FK (2003) Smart homes: past present and future. In: Harper R (ed) Inside the smart home. Springer, London, pp 17–40

Ballantyne GH, Moll F (2003) The da Vinci Telerobotic surgical system: the virtual operative field and tele- presence surgery. Surg Clin North Am 83(6):1293–1304

Castells M (2001) La Galaxia Internet. Areté, Barcelona

Castells M, Cardoso G (eds) (2005) The Network Society: From Knowledge to Policy. Johns Hopkins Center for Transatlantic Relations, Washington, DC. pp 3–15

Castilla A, Díaz JA (2008) La sociedad de la información en España. In: del Campo S, Tezanos JF (eds) España Siglo XXI. La Sociedad, Biblioteca Nueva, Madrid, pp 583–623

Díaz JA (2009) Juventud y TIC: usuarios y suministradores de información en la sociedad del conocimiento. In: Tezanos JF (ed) Juventud y Exclusión Social. Editorial Sistema, Madrid pp 447–470

Díaz JA (2010) El hogar inteligente del futuro: tendencias de cambio en las tecnologías de la información en el entorno doméstico. In: Tezanos JF (ed) Incertidumbres, retos y potencialidades del Siglo XXI: Grandes tendencias Internacionales. Editorial Sistema, Madrid

Drexler E, Peterson C, Pergamit G (1991) Unbounding the future: the nanotechnology revolution. Atlanta Book Company

Forge S, Blackman C (2010) A helping hand for Europe: the competitive outlook for the EU robotics industry, EUR 24600. In: Bogdanowicz M, Desruelle P, JRC IPTS, Sevilla, p 15

Garson DG (2006) Public information technology and E-governance: managing the virtual state. Jones & Bartlett Pub, Massachusetts, pp 12–13

Grupo de Estudio sobre Tendencias Sociales (2002). Estudio Delphi sobre tendencias científico-tecnológicas 2002. Editorial Sistema, Madrid

Grupo de Estudio sobre Tendencias Sociales (2005) Estudio Delphi sobre tendencias científico-tecnológicas 2005. Editorial Sistema, Madrid

Grupo de Estudio sobre Tendencias Sociales (2011) Estudio Delphi sobre tendencias científico-tecnológicas 2011. Editorial Sistema, Madrid

Himanen P (2005) Challenges of the Global Information Society. In Castells M, Cardoso G (eds) o.c., p 341

Hosseini SM, Najarian S, Motaghinasab S, Dargahi J (2006) Detection of Tumours Using a Computational Tactile Sensing Approach. Int J Med Rob Comput Assist Surg 2(4):333–340

Krebs HI, Volpe BT, Williams D, Celestino J, Charles SK, Lynch D, Hogan N (2007) Robot-Aided Neurorehabilitation: a robot for wrist rehabilitation. IEEE Trans Neural Syst Rehabil Eng 15(3):327–335

Lum MJH, Friedman DCW, Sankaranarayanan G, King H, Fodero K, Leuschkeand R et al (2009) The raven: design and validation of a telesurgery system. Int J Rob Res 28(9):1183–1197

Marescaux J, Leroy J, Gagner M, Rubino F, Mutter D, Vix M et al (2001) Transatlantic robot-assisted telesurgery. Nature 27:379–380

Mozer MC (2005) Lessons from an adaptive house. In: Cook D, Das R (eds) Smart environments: Technologies, protocols, and applications. Wiley & Sons, Hoboken, pp 273–294

Mulgan G (2005) Reshaping the State and its Relationship with Citizens: the Short, Medium and Long-term Potential of ICT's. In: Castells M and Cardoso G (eds.), *o.c*, p 237

Najarian S, Fallahnejad M, Afshari E (2011) Advances in Medical Robotic Systems with Specific Applications in Surgery—A Review. J Med Eng Technol 35(1):19–33

National Institute of Science and Technology Policy (2001) The Seventh Technology Foresight. Future Technology in Japan toward the Year 2030. NISPET REPORT 71

Norris P (2001) Digital divide. Civil engagement, information poverty and the Internet world wide. Cambridge University Press, Cambridge

Pérez C (2000) Cambio de paradigma y rol de la tecnología en el desarrollo. Conference La ciencia y la tecnología en la construcción del futuro del país. MCT, Caracas, pp 1–2. http://www.carlotaperez.org/Articulos/CP-Foro-MCT.pdf

Rogozea L, Leasu F, Repanovici A, Baritz M (2010) Ethics, robotics and medicine development. In: Proceedings of the 9th WSEAS international conference of Signal processing, robotics and automation, Wisconsin

Servon L (2002) Bridging the digital divide. Techology, community and public policy. Blackwell Publishing, London

Tapscott D (1995) Digital Economy. McGraw-Hill, New York

Tapscott D, Williams AD (2007) Wikinomics. How mass collaboration changes everthing. Portfolio, Penguin Books, New York

Tezanos JF, Bordas J (2000) Estudio Delphi sobre la casa del futuro. Editorial Sistema, Madrid

Tezanos JF, Manuel Montero JM, Díaz JA (eds) (1997) Tendencias de futuro en la Sociedad Española. Editorial Sistema, Madrid

Tezanos JF et al (2005) Estudio Delphi sobre Tendencias Científicos-Tecnológicas. GETS, Editorial Sistema, Madrid

Tezanos JF (ed) (2000) Escenarios del Nuevo Siglo. Editorial Sistema, Madrid, pp 256–317

The Academy of Medical Sciences (2012) Human enhancement and the future of work. Report from a joint workshop hosted by the Academy of Medical Sciences, the British Academy, the Royal Academy of Engineering and the Royal Society. The Academy of Medical Sciences, October

Toffler A (1984) La Tercera Ola. Plaza & Janés Editores, Barcelona, p 231

Villorte N, Glauser D, Flury P, Burckhardt CW (1992) Conception of Stereotactic Instruments for the Neurosurgical Robot Minerva. 14th Annual International Conference of the IEEE Engineers in Medicine and Biology Society, Paris, 29 October–1 November

Westebring van der Putten EP, Goossens RH, Jakimowicz JJ, Dankelman J (2008) Haptics in Minimally Invasive Surgery—A Review. Minimally Invasive Therapy and Allied Technology 17(1):3–16

Chapter 8
Dependency, Social Work, and Advanced Automation

Yolanda M. de la Fuente Robles and Eva Sotomayor Morales

8.1 Introduction

Attending to situations of dependency, which is understood as "a state, of a permanent nature, in which people who, for reasons of age, illness or disability, linked to the lack or loss of physical, mental, intellectual or sensory autonomy, find themselves in need of the attention of another person or other people, or of substantial help in order to perform the basic activities of daily life or, in the case of people with intellectual disability or mental illness, in need of other support for their personal autonomy" has become an unavoidable challenge for the public authorities, because it requires a firm and sustained response which is adapted to our current model of society (Benítez et al. 2009a).

With attention to dependency by means of information and communication technologies (ICTs) which are implemented by professional social workers, a new stage is being shaped for the social services and for the professionals, who are at last being transformed into generators of quality resources which are designed in a customized manner, all of which responds to the need for attention to situations of dependency and to the promotion of personal autonomy, quality of life, and equality of opportunities.

The ICTs are bursting in transversally as facilitators of accessibility, and of the participation of all, in the exercise of their rights. If we confine our attention to those in situations of dependency, the ICTs put the emphasis, not on aspects relating to the lack of ability, but on those skills and abilities which *can* be developed, thus acting as powerful tools which provide access to equality, and the consequent improvement in quality of life. The ICTs must be an opportunity to advance in social cohesion, and a real source of opportunities.

Y. M. de la Fuente Robles · E. Sotomayor Morales (✉)
Departament of Psicology, University of Jaén, Jaén, Spain
e-mail: esotoma@ujaen.es

Y. M. de la Fuente Robles
e-mail: ymfuente@ujaen.es

A. López Peláez (ed.), *The Robotics Divide*,
DOI: 10.1007/978-1-4471-5358-0_8, © Springer-Verlag London 2014

In short, the factors which contribute the most to the social exclusion of people with disabilities, according to the study "Disability and Social Exclusion in the European Union: time for change, tools for change" (2003), are the following, in order of importance: "the lack of, or limited access to, goods and services," "the stigmatization of people with disabilities," "the lack of appropriate training, the lack of specialized services," "the inappropriateness of the educational systems," "the lack of economic policies to compensate for the extra spending resulting from disability," "the structure of the subsidy system," and finally, "living in institutions."

The solution to many of these handicaps, both those which affect people in dependency situations and those affecting the people who care for them, can be found in the ICTs, because these can be a factor in union, communication, training and innovation in as much as they fall within the emerging paradigms of "environmental intelligence": ubiquity, because they can accompany users wherever they are (at home, in school, in some means of transportation, in hospital, moving in the street, etc.); invisibility, because they can pass unnoticed in the physical environment; and adaptability, because they can be adapted to the preferences of each person (De la Fuente and Sotomayor 2009).

One of the most innovative and most representative cases of "Dependency, Social Work and Advanced Automation" is the use of ICTs as a linking bridge between the social workers, the people in dependency situations and their carers, and we shall devote this chapter to that link.

8.2 Current Situation

Within the framework of the European Union and the countries forming the OECD, there is frequent political debate about the protection of dependency and long-term care (LTC), which is manifested in the form of public policies of universal cover, through the Social Security or through attendance, expenditure on which had reached as much as 3 % of GDP in countries such as Denmark, and a weighted average expenditure of 1.3 % in countries which provided it in 2003 (IMSERSO 2005b).

These policies make it possible to provide services, pecuniary or in kind, to the person dependent on attention in their own home, in specialized centers or in a combination of these, thereby directly ensuring the quality of the care and indirect support for the carer.

If we consider the support which is given to the carers, we can see different options of direct support within the system of protection of dependency situations, in their different forms (Table 8.1).

As can be observed, the Nordic countries, in accordance with their welfare state model, provide ample cover, whereas there are no resources for informal support, due to its scant establishment. With regard to social security, the German model stands out because of its wide network of resources for carers, which provides not only training, but also rest periods and other relevant aspects.

Table 8.1 Bibliography table 8.1

Zone	Dependency model	Support resources informal carer
Nordic countries	Universal	No tradition of informal support, institutionalization
Holland	Universal	Financial compensation for the family carer
Germany	Social security	Accident insurance and contributions cover; temporary replacement (maximum 4 weeks); technical helps and adaptation of the home; residential attendance in situations of crisis or for rest; free training courses
Austria	Social security	Advice service; social security for carer families
Luxembourg	Social security	Accident insurance; pension contribution; four weeks to hire professional services; adaptation of the home
France	Social security	Training and advice
The United Kingdom	Attendance	Welfare benefit for the carer with a low income; training programs; possibility of vacation period
Australia	Attendance	Welfare benefits for carers
USA	Attendance	Certain measures for disabled people and in favor of carers
Japan	Universal Insurance (those over 40); attendance for those without resources.	Assistance for carers
Switzerland	Social security (contributive right)	Assistance for carers of medium-grade dependants
Spain	Universal	Social security carer families; supporting welfare benefit (exceptional); measures for training, information and qualification; measures for rest

Source Own preparation from bibliography consulted (IMSERSO 2005a; Benítez et al. 2009b)

The countries where informal support is most relevant are those in which protection for dependent situations has not developed or is residual, as is the case in the countries of Southern Europe. This was the Spanish case until the arrival of Law 39/2006, of 14 December, for the Promotion of Personal Autonomy and the Care of People in Situations of Dependency, with which the social services acquired a universal character in contrast to their original idea of attendance.

But in order to unify all of the foregoing, the European Union laid down a series of common objectives to be achieved by the member states, among which, according to Benítez Bermejo et al. (2009), were as follows:

- The need to devise new ways of supporting the family or the informal carers;
- The integration of informal carers into the labor market, and a better development of their working conditions.

8.2.1 The Carers

The studies and analyses which are currently being carried out internationally define LTC as a variety of healthcare and social services that are provided over a long period to people who need permanent care because of physical or mental disability (European Commission 2008).

The *carer* has been defined as "the person who attends on or cares for another who is affected by any type of disability, handicap or invalidity which impedes or prevents the normal performance of the activities of their daily lives or social relationships" (Flórez Lozano et al. 1997).

Starting from this general definition of carers, among the direct carers, it is necessary to differentiate between *informal carers* and *formal carers*.

Informal carers are people who care for or help non-autonomous people with long-term physical or mental illnesses or who have gerontological problems, generally in their homes, and are rarely remunerated. These carers can be relatives of the dependent person or they can be volunteers of other types. When they have to deal with serious problems, often in difficult circumstances, the carers have a particular need for attention and support from national, regional or local authorities, who have to help them in their work. In general, in order to carry out their task efficiently and without complications, they need a combination of supporting services which is appropriate to the provision of social and medical assistance to the dependent person, and also financial support (including, for example, payment of social security contributions) and flexible conditions in their own professional work.

By contrast, formal (professional) carers provide long-term assistance as employees of a public or private body (or company), whether in medically equipped institutions or at home. The professional care can be entrusted to, for example, nurses or nursing auxiliaries with or without qualifications.[1]

As the boundaries of informal care are imprecise, an indispensable condition of its definition is considered to be the absence of a contract or previously agreed working relationship between the parties, although it can occasionally be remunerated or compensated (Rogero García 2010, p. 40).

The typical carer in Spain, according to the 2008 Survey of Disability, Personal Autonomy and Dependency Situations, is: a woman(76.3 %), between 45 and 65 years old, with a certificate of primary studies or equivalent, whose economic activity is either housework or employment, who co-habits with the dependent person (79.3 %) and devotes more than 8 h a day to caring. With regard to the kinship with the person cared for, the carers are usually daughters (38.8 %), the spouse (21.8 %) or the sons (10.1 %), according to the IMSERSO-CIS (2006), but these data only relate to carers of elderly people.

---

[1] More information from: http://ec.europa.eu/health-eu/care_for_me/carers/index_es.htm.

Langa and Martínez (2009) add to this profile: "the way in which inequality of resources generates inequality of opportunities and methods of becoming a provider of care to dependent families."

If we compare this profile of the non-professional carer with that found in other European countries (Greece, Italy, the United Kingdom, Poland, Sweden, and Germany), in the case of the care of elderly people between 50 and 65 years old, the carers are normally women who co-habit with the person cared for. However, with regard to other variables such as kinship, the work situation or the number of hours devoted to the care, the results are different in each country (Balducci et al. 2008).

We should not forget that informal support is very important in the whole of Europe. The task is assumed mainly by women, who accept responsibility for approximately 2/3 of the informal support (EUROFAMCARE).[2] The only case in which attention is given by similar percentages of men and women is that of elderly people who take responsibility for their spouses; however, there are very significant differences in how and to what extent the different European societies respond to the problem presented by dependency. We can group them around two axes: The social expectations with regard to the role of the family in the care of its elderly members and the level of services offered by the public sector. The first axis is closely related to the social role of women and their participation in the labor market, and the second can be approached quantitatively through the public spending on LTC. The two dimensions are interconnected, the countries in which the family assumes most of the caring being those where the public services are the least developed, and vice versa. It is thus the cultural expectations that give rise to the different combinations of formal and informal attention, and to the different levels of recognition of the task carried out by informal carers (Mestheneous and Triatafillou 2005). In general, we can establish two patterns of attention to dependency: that of the North and that of the South, which differ in three areas: The dependant's place of residence, that is to say, whether they live in the same home as the main carer; the extent of the female participation in the labor market and, especially, of the participation of the informal carers; and the level of development of the formal services. The Northern pattern is characterized by the

---

[2] EUROFAMCARE: An international research project, financed by the 5th Framework Program of the European Union, whose purpose is to carry out a review, at European level, of the situation of carer-relatives of elderly people through the creation of a working group with members from six countries: Germany, Greece, Italy, Poland, Sweden, and the United Kingdom. Each of these countries will gather data from approximately 1,000 interviews. One of the aspects that will be analyzed is the situation of carer-relatives of elderly people in relation to the existence of, familiarity with, and availability, use and acceptance of the different support services. Eurofamcare studies aspects such as current demographic trends; the legal obligations of carers and the role of the state; the role of the carer and social attitudes; the "work" of caring; public investment in the care of the elderly; the professional carers; and finally, Eurofamcare will present some conclusions and some policy recommendations. This project is intended to bring about changes in management processes at various levels, with a view to promoting social policies for the benefit of carer-relatives of elderly people.

non-cohabiting of carer and dependant, by a greater participation in work and by the wide range of public services (all in comparison with the characteristics of the Southern countries). In general, the Scandinavian system follows the Northern pattern, and the Mediterranean countries, the Southern model. The Continental countries would be in between the two (Jiménez-Martín 2007).

8.2.1.1 Informal Care

Of Spanish carers, 54.4 % was stated that there have been effects on their working lives and on their finances, and 63.7 % have reduced their leisure time, as a result of attending on a disabled person. The carers feel that they have difficulty in carrying out their tasks, and they also consider that their health and their personal lives are deteriorating: They stress that they feel "worn out" and tired (INE 2009, p. 4).

The Scale of Carers' Quality of Life, hereafter called the SCQL, prepared by Glozman et al. (1998), allows us to evaluate, qualitatively and quantitatively, the main activities in the life of the carer, arranged into three groups: (a) the carer's professional activity; (b) social and leisure activities; (c) the carer's responsibilities in the daily care of the dependant (Jiménez-Martín 2007).

The decision to become an informal carer involves a series of consequences and changes in the person's life which, according to Rogero García (2010: 58), if we confine our attention to the negative consequences, include: *damage to health,* both physical (tiredness, back problems, etc.) and psychological (stress, "burn-out," depression, etc.), *costs to the economy,* both indirect costs (contributions, revenue, productivity at work) and direct costs (handling expenses, technical assistance, payments for services) and *damage to social relationships,* both in the family (deterioration or reduction in the number of family relationships) and in the extra-family (fewer relationships, less social participation).

There are three important factors underlying the concept of the carer's responsibility: First, the *impact of the care provided,* which alludes to the confronting of a new family situation; secondly, the *interpersonal responsibility*; and thirdly, the *expectations of auto efficiency,* in other words, the ability to respond to the needs of the person cared for. We should not forget that "caring is one of the situations most intrinsically representative of the generation of chronic stress" (Montorio et al. 1998).

When one analyzes the way of confronting this type of situation, it is interesting to note the application of Losada et al. (2006: 36) of the models of stress and confrontation of Lazarus and Folkman (1984) to the situation of care. These highlight the influence of the contextual variables (gender, age, health, and kinship with the person cared for); the demands related to the situation of the person cared for (objective stressors and how they are perceived by the carer); strategies to confront these (which are, according to the authors, the variables mediating between the evaluation of these stressors and their consequences); and finally, the consequences of the demands of the person cared for. The assessment of the cost of care in each situation will therefore depend on those variables.

Mnich and Balducci (2006) studied six different profiles of carers or different care situations: burdened unemployed, burdened employed, burdened wives, burdened husbands, tense but unburdened carer, and carer without burden or stress. The study was carried out in Greece, Italy, the United Kingdom, Poland, Sweden, and Germany. The results showed more carers with heavy burdens in Greece and Italy and fewer in Poland and the United Kingdom, the attention of "burdened" wives being noteworthy in Sweden. In Germany, however, none of these profiles was particularly noticeable. One interesting fact that emerged was that, in all the countries, there were more wives than husbands who were carers (Mnich and Balducci 2006). In this connection, the perception of each person's own health is also different in each country, being "very good" (80 %) in Sweden and "reasonable" in Poland and Italy.[3]

8.3 Social Work and the ICTs as a Tool for Carers

8.3.1 Social Work and Family Intervention in Dependency Situations

The appearance of a social worker as a professional, intervening in a health process, is almost always linked to the scarcity of financial resources or to conflictive situations surrounding the patient. However, dependency is aggressive toward the environment, regardless of the financial resources available to it.

The person has, until now, been integrated into a network of relationships which now has to change to adapt itself to the new situation. A genuinely complete attention will have to take into account all of the relationships between the different elements, as well as their characteristics. The intervention of a social worker permits a veritable total focus on situations of dependency.

The social worker carries out a social and family evaluation of the dependent person to help in the designing of an intervention plan. In this evaluation, the following are all considered: structural aspects of the family, the distribution of roles, previous relationships, the financial situation, the geographic environment and the facilities, the social/family perception of the situation, the social and family functioning of the dependent person, the habits, the level of stability in the environment, the impact on the life expectancies of the carers, etc. The relative position of the factors which have a bearing on the situation and their weight in the whole vary in each case, and at each moment, in one and the same situation.

---

[3] Mckee K, Lamura G, Prouskas C, Öberg B, Krevers B, Spazzafumo L et al. The COPE Index—a first stage assessment of negative impact, positive value and quality of support of caregiving in informal carers for older people.

The social worker knows the situation from the beginning of the process and collaborates in the evaluation and in the overall "treatment" of the dependent person and of their family system, so their contribution has a therapeutic content and aims to be preventive as well as attentive. Social workers intervene in the "total treatment" so that the family system finds its equilibrium and re-adjusts; in other words, they accompany the family in the quest for solutions, in order to cope with the objective and subjective needs presented by the situation (the professional is not the one who decides which requirements have to be satisfied: It is the family itself which must find the appropriate response to these, with appropriate professional support which serves as orientation). Those needs will depend, on the one hand, on the way the illness itself develops, and on the other hand, on how the family is structured and how it functions. It follows that, just as there is no pattern of illness, neither is there any model of action to be taken.

The aim of social work is to facilitate growth and adaptive change in the face of new situations which de-stabilize and have negative repercussions on the family system; but it is also designed to prevent imbalance and upset in the family, in matters of orientation and decision which require technical advice. So, it is a matter of helping the subjects and their families to achieve more autonomy, when that has been reduced by external or internal constraints, or by the interaction of both.[4]

Family Social Work is a specialized form of social work which takes as its scope of work the family and its family relationships, and considers the context into which these are inserted. By these means, it aims to address the psychosocial problems which affect the family group, generating a process of directing help, which seeks to foster and activate both the resources of the people themselves and those of the family and the social networks (Donoso and Saldias 1998).[5]

8.3.1.1 ICT Resources and the Carers

Until now, social workers have traditionally been considered as directing their efforts toward the list of resources, but it is necessary to go further, to the participation of the people on whom the intervention is performed, and here, the most appropriate tool is the one related to the ICTs, because it is free and autonomous.

There have been experiments with this in different countries, but they were isolated, and social workers barely know of their existence. This is the case of the *Proyecto Pecujici* (www.pecujici.cz) in the Czech Republic, which offers online training for informal carers of elderly people with chronic illnesses.

---

[4] More information from: http://www.famma.org/rokdownloads/Publicaciones/2008-guia-para-personas-cuidadoras.pdf.

[5] Article available on: http://www.ts.ucr.ac.cr/binarios/congresos/reg/slets/slets-016-059.pdf.

Another initiative is the *CLIC* centers in France (Local Centers for Information and Co-ordination) in Gerontology, where information is shared about the evaluation and definition of services for elderly people and their families.

In the *Programa Sendian*, in Spain (Guipúzcoa), different financial, social, educational, and psychological services are provided for informal carers; the activities organized by the voluntary workers in this program cover training, psychological support, alleviating the workload, and technical help. From the financial point of view, fiscal helps and benefits are envisaged.

The *Punto Insieme* initiative, in Tuscany (Italy), is a service of advice and information where one can ask for help to improve the care of sick relatives.

In Slovakia, too, a great effort has been made to develop mechanisms of cooperation between families with members who need LTC and town councils, social services, information centers, doctors and employment services.

In the TioHundra district of Sweden, they have created a joint organization for health, medical services, and informal care, with the aim of improving attendance, planning and care management from the medical point of view, but also for the informal carers.

Finally, in Greece, the Alzheimer charity provides wide-ranging programs of information, advice, periods in old people's homes, and training for carers of those with Alzheimer's disease.[6]

It is important to remember that family carers have no training in medicine, do not receive financial remuneration, and fulfill this role without following any regulation of working hours or of procedure. These people, the majority of whom are women, spend a large part of the day with the person whom they are caring for and in many cases co-habit with that person (Giraldo et al. 2005).

For all these reasons, the ICTs should be considered as a vehicle for a formative and actively participative response to carers' needs in four fields:

1. *The need for information about illness*, its care and complications: what could possibly happen and how to act, how to anticipate events so as to be prepared to confront them. There is also a need for information about the available resources and assistance.
2. *The need for training* in the development of communication skills, in techniques for dealing with stress, and in caring for the patient and for themselves.
3. *The need for emotional support*, to receive affection and understanding from people and one's social circle.
4. *The need to increase social support*.[7]

---

[6] More information from: http://www1.unavarra.es/digitalAssets/158/158808_5_Carretero_PoliticasSociosanitariasSostenibles.pdf.

[7] The needs of the main carers: http://www.centrodesaluddebollullos.es/Centrodesalud/Enfermeria/Documentacion%20Distrito/Documentos/Cuidadoras/ElSistemaInvisibledelosCuidados.PDF.

Unlike professional care workers, those who are not professionals lack a fixed timetable, a salary, a guaranteed vacation period, unemployment benefit and a service for the prevention of risks at work, among other benefits. It must be added that becoming the carer of a relative is an event of vital importance which begins without any preparation to exercise the role, and it is interesting to reflect on the ability of these people to provide that attention with prior training.

However, intervention with, and monitoring of, carers does not occur in a specific manner in dependency protection systems and is not an essential axis of policy in the countries which have developed those systems of protection.

The current project therefore aims to study a possible tool, or "new way" by which to support carers, based on the opportunities which the ICTs can provide for the empowerment of the carer as a result of the alternative world offered by the Internet, because:

"The human being, who through language and writing was always the artificer of communications, today communicates exponentially, using the new technologies" (Raya and Santolaya 2009). The Internet is a world in which the potential of the processes of intervention developed through social work with different groups experiencing difficulties is asserted.

Authors such as Mata et al. (2009) point out the opportunities offered by the Internet: in administration, health, training, leisure, etc., and add that we are working toward the achievement of a Society of Information, without barriers or distances, and with e-access and e-inclusion fully available.

In this connection, workers in the Fundació Surt[8] point out that the ICTs are useful tools in our personal, social, and working lives. They highlight a long list of services, resources, and technologies which can be found on the Internet, noteworthy among which are: obtaining information, consulting electronic communication media, contributing information to online communication media, instantaneous communication (messaging, "chat," telephone calls, sending of "SMSs"), television by Internet, procedures and requests for the public administrations, job-hunting, electronic mail, training, electronic shopping, etc. This foundation also mentions the risk of social exclusion if, in today's world, the ICTs are not used, as well as the importance for women of using the ICTs.

This reality of the use or non-use of the ICTs can be related to what Marc Prensky[9] calls *digital natives and immigrants*, referring to the "digital divide" as the confrontation between two generations: the one that was born without digital devices (the digital immigrants) and the one which cannot now comprehend the world without them (the digital natives). The first have to adapt to this environment (like all immigrants) and make only a secondary use of the ICTs, whereas the digital natives use them as a primary resource.

---

[8] Fundación Surt: http://www.surt.org/zonatics/?page_id=6.

[9] Complementary information from: http://aprenderapensar.net/2009/05/18/nativos-digitales-vs-inmigrantes-digitales/.

For their part, Keeble and Loader (2001) point out that the IT community provides, through the computer, opportunities such as mediation/social support, self-help, "chat rooms," interchange of knowledge.

If we emphasize the support functions of the ICT initiatives for carers, we can find the response to many of the needs mentioned above, such as the need for different kinds of support, that for the dependent elderly person, and that for the quality of attention of the carer, or for their quality of life and participation (Schmidt et al. 2011: 11).

One of the barriers that is sometimes encountered when using this resource is the "digital gender divide": among men aged between 16 and 74, 63.4 % have used the Internet at some time, whereas among women in the same age range, the percentage is 56.2 %. This divide is not significant in the age group 16–24 and relates, among other things, to access to and use of the Internet.[10]

On the other hand, according to the *Instituto Nacional de Estadística* (National Statistics Institute), in 2010 59.1 % of Spanish households had access to the Internet, and 58.4 % of the population aged between 16 and 74 used it frequently. The reasons for not having Internet at home are: because they do not need it (56.3 %), because they have little knowledge about how to use it (28.2 %), because the cost of connection is too high (24.8 %), or because the cost of the equipment is too high (23 %). The same report states that Spain is situated within the European average, the same as countries such as Ireland or the Czech Republic, the countries of Northern Europe standing out above the average, and those in the Mediterranean area, together with the recent members of the EU, being below it.

If we now consider the carers, in 2005 a study by Blackburn et al. (Blackburn et al. 2005) of carers and the digital divide in the United Kingdom highlighted among its results the fact that 50 % of the carers had previously used the Internet, and that of that group, 61 % had done so once a week or more frequently. In this connection, the amount of use of the Internet was related to the age of the carer, the work situation, the tenure of the home, and the number of hours per week devoted to caring. There was a significant connection between the frequency with which the Internet was used and the carer's age, sex, work situation, and number of hours devoted to caring. This study suggests that, currently, a large numbers of carers cannot be Internet users because of their age, sex, socio-economic situation, or family responsibilities.

In addition, if we study the prospects of the carers' using the Internet, Read and Blackburn (2005) identified as the main barriers to this: the lack of access to the equipment and to the Internet, difficulties in using the equipment and systems, the cost, limitations of time, and lack of interest and ability. The key benefits identified included convenience, flexibility, speed, and the extent of the information

---

[10] Official data of the *Ministerio de Industria, Turismo y Comercio* (Ministry of Industry, Tourism and Trade 2010).

available, whereas problems with the equipment and systems, and lack of time, were the main obstacles to the efficient use of the Internet.

However, bodies such as the European Commission (2008) promote the use of ICT to support dependency, but not as a specific support for the carers.

One must therefore consider, as pointed out by the *Grupo de Género en la Sociedad de la Información* (OSSIC) (Gender Group in the Information Society), that reducing the digital gender divide implies not only increasing the number of Internet users, but also reviewing the content and analyzing the situation of women, their needs, priorities, and wishes. The same group asserts that the ICTs represent an opportunity for all women, and offers various possibilities and uses: contact with other people, creating relational and interest groups, job-hunting, training, creating networks and, especially, they provide a space for generating and finding information in an alternative manner, which overcomes difficult aspects such as the lack of time (double or triple working day), the invisibility of womens' knowledge and creativity or the existence of misogynous content or of material designed without taking account of the woman.

An active participation in the information society can become an essential element in the development of one's independence, autonomy, and creativity, enabling new social networks to be created and maintained and avoiding isolation from society. It also facilitates access to services (health or cultural services, etc.) which results, in short, in a better quality of life (Gracia and Guerrero 2007).

Hence, the importance of taking account of the implications which the digital divide can have for this section of the population—by creating barriers which deprive a large number of elderly people of the numerous benefits brought by the use of the ICTs (OCDE 2003).

The inequality in the use of the new technologies and the benefits derived from such use can be due to the following causes:

- Lack of awareness: Many elderly people[11] are unaware of the usefulness of certain technological devices and their applications.
- Complexity of use: They are worried by the idea of entering into a world which is too complex.
- The feeling of a different era: Among the elderly people who are introduced to the world of the new technologies, there is a feeling that they belong to another era, not that of these new instruments.
- Exclusion from the consumer society: Currently, technological products are the ones which are advertised the most. However, elderly people rarely appear in those advertisements. The consequence is that, being products which, apparently, are not intended for the elderly, they produce a feeling of indifference and a lack of interest.[12]

---

[11] The resource must be adapted to the profile of each carer, most of whom are elderly.

[12] The complete article may be found on: http://blog.catedratelefonica.deusto.es/tecnologia-y-personas-mayores/.

Table 8.2 Entities with ICT programs for the elderly

Entity	Program	Activities
Fundación la Caixa	Cibercaixas	Face-to-face workshops
Obra Social Caja España		Face-to-face and online courses
Obra Social Caixa Galicia	Ciberaulas	Courses in "digital literacy"
Fundación Esplai	Red Conecta	Teletraining platform based on other users' experiences
Fundación Insula Barataria	Plataforma teleformación	Face-to-face and online training
		Access centers

Source Own preparation

8.3.1.2 Benefits of the Use of ICTs

The use of ICTs gives rise to numerous benefits, because it *cultivates and extends the network of relationships* and the use of mobile telephony, while tools such as electronic mail and instantaneous messaging *speed up communication* with other people, strengthen independence—because they allow immediate access to information about resources which can serve as supports in daily life—and *keep their users more active and healthy*. In Spain, there is a group of bodies that fight to eradicate the digital divide by offering free training, advice and motivation to elderly people who decide to take the plunge into the Internet, and to show them its usefulness and benefits in the development of their lives.

Many foundations and non-profit-making entities dedicate part of their efforts to the creation of programs and workshops which help them (Table 8.2).

So, if we consider person/surroundings relationships, these can be analyzed as a complex network of transactions. One of the conceptual definitions of this process is based on the analysis of compatibility. There is compatibility when a correspondence is noticed between people (their needs, their intentions) and the environment (their opportunities, their requirements for action). Incompatibility has its origin both in the inappropriate operation of the environment (its restrictive capacity, its informative poverty, repetitiveness, etc.) and in inappropriate dispositions of people (internal conflicts, lack of interest, lack of personal resources for planning or organizing their own action, etc.). If we take these premises into account, we can list three types of environments: the supportive (whose characteristics favor the development of people's plans and goals), the controllable (in which the person can develop their plans and goals even if they find this difficult), and the restorative (in which the person can recover from the unwanted pressure of environmental stimuli) (Fernández-Ballesteros and Corraliza 2000).

In addition to promote the use and knowledge of ICTs, through the design of the different devices, a conceptual change is taking place in order to make them more accessible and easier to use ("design for all"). To guarantee accessibility, measures such as the W3C, WAI are considered among others, and the different legislative directives about equality of opportunities, non-discrimination, and universal accessibility are complied with at all levels.

This is the compass which must guide the specific actions of the public powers in moving from the current situation of the carers toward a real and effective participation in the society to which they belong as seekers of services and beneficiaries of personalized and instantaneous attention, thus breaking with the hostile environments which marginalize and exclude them in their daily lives.

It is therefore essential to open channels to personal and collective initiative, and it is also necessary to guarantee accessibility, so that every kind of barrier can be dismantled.

8.4 The New Social Work in the Context of the Society of Knowledge and of Robotics in Services

In periods of crisis, the professions which respond to social needs are confronted with new situations which, if correctly monitored, can present new opportunities for development.

Social work today falls within a social, economic, technological, and cultural context which can be considered as a stage of re-organization of the profession, which now presents new characteristics that make greater demands on the training of social workers. The ICTs bring with them new opportunities in the different forms of intervention: as regards *direct intervention,* when it takes place in various environments and includes a series of professional actions on the part of the social worker which, to achieve their objective, require personal contact between the professional and the person, family or group involved, so that the relationship which is established between the social worker and the "client system" (individuals, family, small group) is a significant factor in the change of situation. In this model, the ICTs would generate a continuous virtual support between the social worker and the carer, giving rise to a new era for the community social services and primary care services, since the carers who are included in a particular network can be informed, orientated, and advised in all situations which involve both the dependent person and themselves, thus breaking with isolation and lack of participation.

With regard to *indirect intervention,* when it takes place in different environments and includes activities of study, analysis, systematization, planning, evaluation, co-ordination, and supervision, the ITCs provide a multitude of opportunities for the exercise of the profession in all the professional areas: attendance, preventive, promotional/educational, mediation, planning, evaluating, rehabilitating, and transforming.

As it is not possible to understand the function of prevention without an effort of education to help people and social groups to make use of the opportunities available to them, training them for the same, the case of the ICTs being a clear example of this, social workers increase their knowledge to try to change social practices which create inequality and social injustice, thus affecting the autonomous development of the subjects, contributing to well-being, and enabling the social integration of those individuals and groups who, for personal or social reasons, find themselves socially disadvantaged, this being the case of the carers of dependent people.

That is why the refusal of social workers to think about the implications of the new technologies, for society in general and for their profession in particular, will not prevent their introduction and use, but there will probably be an isolation from the new fields of work, and influence and importance in case of possible interventions will be lost. We find ourselves in the presence of a new tool for work, which can provide us with greater efficiency and the opportunity to create new solutions.[13]

There are many examples of the relation between social needs and the ICTs, and these are found on the interactive and multilingual Web site of the *Association for the Progress of Communications,* which contains the best and most relevant content relating to training in the use of the Internet, specifically developed by and for the charity and development organizations and other groups in the civil society, on *computers in mental health (CIMH),* a complete database of computer-based resources in mental health and IT resources in the area of social well-being, on *Colectivo Virtual Trabajo Social (Social Work Virtual Group),* an independent project developed by Chilean professionals, and which has created a space for opinion and for the professional development of social work on the Internet, on *CTI for Human Services Resource Guide,* which includes IT resources in social services, on *computer use in social services (CUSS),* which contains a lot of interesting information, has a section with software, freeware, shareware, and demos dedicated to social work, on *Digital Opportunity Channel,* a thematic, One World channel dedicated to the new ICTs for sustainable development, on *División digital: Mejorando el acceso a las NTI (Digital divide: improving access to the NITs)*—the OECD has a list of sites and links concerning the digital divide, arranged by country, on *GT II. World Solidarity Web,* an independent working group and meeting point for Hispanics on the Internet, on *human service information technology applications (HUSITA),* which is dedicated to the applications of IT to the social services, on *New technology in the human services,* a European network of entities and people who are worried about IT in the social services which was developed by the Centre for Human Service Technology of the Department of Studies in Social Work of the University of Southampton, on *Social Work Access Network (SWAN),* a reference for seekers of resources in social work, developed by the School of Social Work of the University of South Carolina (USA), on *SWBIB: New Technology in Human Services,* a resource for researchers in social work who are interested in the use of IT, etc.[14]

The European policies for e-Inclusión, and the concepts of e-Accessibility and AAL, have encouraged the participation of everyone in the society of knowledge, as users of the same, especially elderly and/or dependent people, regardless of the

---

[13] Article available on: http://www3.unileon.es/ce/ets/ficheros/informacion/nticts.pdf.

[14] Complementary information on: http://www.tsocialcan.com/centrodedocumentacion/enlacesdeinteres/enlaces/index.php?id_categoria=1.

age, gender, education, or origin of those people. The main objective is that the dependent person should have as much autonomy as possible for as long as possible, ideally in their home environment. The European program AAL appears as a specific program whose principal purpose is the financing of activities which, among other things, prolong the period of their lives which is spent by people in their own environments by increasing their autonomy, self-confidence, and mobility, maintaining their health and capabilities, promoting an improvement in the quality of life of the chronically ill, and increasing resources for the families and the carers. Specifically, the aim is to improve the emergency systems by using innovative products based on ICT—which allows the costs of health systems and social care to be reduced and also allows the industrial exploitation of results which favor adaptation and compatibility among the different regions of Europe.

Many of these initiatives, as well as favoring the inclusion of dependent people, also favor that of their carers, mainly of the so-called informal carers (spouse, father, mother, son, etc.) who find themselves with more help in their daily care, and accompanied in the exercise of their functions by technological tools which facilitate their work.

If, within this universe of attention, we take one more step forward, and we introduce the concept of services robotics for the elderly, they would make possible new technological solutions to help elderly and disabled people and endow their users with a better overall quality of life. Among them, we can highlight, on the one hand, the social and telepresence robots to assist with the care of people, and on the other hand, robotized systems to improve mobility, accessibility, and rehabilitation.

In relation to the former, the social and telepresence robots to assist with the care of people,[15] we can highlight the following:

- *Evolution Robotic* (USA). Robotic system ER1. Requires the mounting of a portable computer on the mobile structure.
- *Fraunhofer IPA* (Germany). Care-O-Bot II, Rob@work, Secur-O-Bot.
- *Gecko Systems* (USA). CareBot, personal robot, useful for the care of children and the elderly.
- *In Touch Health* (USA). Systems for telemedicine which include the telepresence RP-VITA (developed together with iRobot) for use in hospitals, homes, and old people's homes.
- *Mantaro* (USA). The telepresence robot MantaroBot can be used for communication between patients and their relatives or doctors.
- *Risoluta* (Spain). The social robot "Quillo" ("Mate") and the LudoSys system, orientated toward the clinical and educational sectors.

With regard to the latter, the robotized systems to improve mobility, accessibility, and rehabilitation, the following can be highlighted:

---

[15] More information on: http://www.roboticadeservicios.com/robots_ayuda.html.

- *Bluebotics* (Switzerland). Robotic arm ERA-5/1, applicable to domestic environments.
- *BTS* (Italy). Robotized bed Anymov which enables the rehabilitation of the lower members.
- *Cyberdyne* (Japan). A company which developed and markets the exoskeleton Robot Suit HAL.
- *Ekso Bionics* (USA). A company which manufactures and markets exoskeletons.
- *Exact Dynamics* (Holland). Assisting robot manipulator ARM.
- *Hocoma* (Switzerland). A company specialized in developing robotized rehabilitation systems. Among them are the Lokomat system for the lower members, and the Armeo system for the upper members.
- *Kinova* (Canada). The manipulating robot JACO is light, and designed to be mounted, for example on wheelchairs, to be controlled by disabled people.
- *Motion Control* (USA). Prosthetic arm "Utah Artificial Arm."
- *Motorika* (USA). Robotized systems for the rehabilitation of upper members (ReoGo), and lower members (ReoAmbulator).
- *Össur* (USA). Distribute different bionic lower prostheses such as the "Rheo Knee" and the "Proprio Foot."
- *Tek RMD* (Turkey). Robotic mobilization device (RMD) which allows the user to move about either when standing or when seated, inside or outside a building.
- *Tmsuk* (Japan). A robot which helps disabled people. PRE-HOSPITAL CARE ROBOT.
- *Rolland* (Germany). A robotized wheelchair.
- *Victhom Human Bionics* (EEUU). They develop lower prostheses, such as the bionic knee "Power Knee."
- *Yobotics* (USA). Exoskeletal leg RoboWalker, to facilitate rehabilitation.

Among all these solutions, we can highlight a robot which helps and provides company for elderly people, and equipment which detects human activity by means of analysis using advanced video cameras incorporating movement sensors and navigation, such as the Senscam of the University of Dublin in Ireland.

In the case of Carnegie Mellon of the University of Utah, its solutions are able to combine information about the person's wandering in their environment with physiological parameters such as heartbeat, respiratory rhythm, body temperature, and information about the person's posture. Also, noteworthy are solutions for mobile devices which enable the identification of people's activities.

8.5 Conclusions

Because of everything that is explained above, it has become necessary to drive forward the full integration of carers and cared for, and the resources which are intended for them must be participative, dynamic, and innovative. One opportunity in this field is, without any doubt, advanced automation, and its introduction through automated or robotized services systems.

From the point of view of the advances in these technologies, and in the adoption of innovative products and services, a real advance will occur when it becomes possible to measure other aspects which are crucial to innovation, such as the complexity of installation, the non-intrusiveness for the users, and the acceptability of the systems for elderly people.

Robotizing services aims to make daily activities easier, so it is essential to continue with the research so that all the advances can contribute, in the medium term, to the effective availability of robots, which will help considerably in improving people's circumstances and in providing them with a longer and longer life expectancy.

The ultimate aim must be to determine which are the best alternatives on which to base future developments of services and applications in those areas of life which are assisted by the surroundings, by eInclusion and by eHealth.

References

Balducci C, Mnich E, Mckee K, Lamura G, Beckmann A, Krevers B et al (2008) Negative impact and positive value in caregiving: validation of the cope index in a six-country sample of carers. Gerontologist 48(3):276–286

Benítez B et al (2009) El sistema estatal de atención sociosanitaria en el ámbito de la dependencia. El caso español y otros modelos de referencia. IBV, CUIDA

Benítez E, Poveda R, Bollaín C, Porcar R, Sánchez J, Prat J et al (2009a) El sistema estatal de atención sociosanitaria en el ámbito de la dependencia. El caso español y otros modelos de referencia. Instituto Biomecánico de Valencia, Valencia

Benítez E, Poveda R, Bollaín C, Porcar R, Sanchez J, Prat J et al (2009b) El sistema estatal de atención sociosanitaria en el ámbito de la dependencia. El caso español y otros modelos de referencia. IBV and CVIDA

Blackburn C, Read J, Hughes N (2005) Carers and the digital divide: factors affecting Internet use among carers in the UK. Health Soc Care Community 13:201–210

Comisión Europea (2008) Long term care in European Union. Dirección General de Empleo, Asuntos Sociales e Igualdad de Oportunidades. Comisión Europea

De la Fuente Y, Sotomayor E (2009) Las Tecnologías de la Información y Comunicación (TIC) como Instrumento de Ejercicio de Derechos. Revista Tabula Rasa, Núm. 10, enero-junio. Universidad Colegio Mayor de Cundinamarca, Colombia, pp 359–373

Fernández-Ballesteros R, Corraliza R (2000) Ambiente y Vejez. Gerontología Social. Pirámide, Madrid

Flórez Lozano JA et al (1997) Psicopatología de los cuidadores habituales de ancianos. Revista Departamento de Medicina, Universidad de Oviedo, Barcelona

Giraldo C, Franco G, Correa L, Salazar M, Tamayo A (2005) Cuidadores familiares de ancianos: quiénes son y cómo asumen este rol. Revista de la Facultad Nacional de Salud Pública 23(2):7–15

Glozman et al (1998) Scale of quality of life of care-givers. Springer

Gracia E, Guerrero J (2007) Brecha digital y calidad de vida de las personas mayores. Available via http://www.mundointernet.es/IMG/pdf/ponencia140.pdf. Accessed 2 Jan 2011

IMSERSO (2005a) Libro Blanco de Atención a las Personas en Situación de Dependencia en España. Ministerio de Sanidad y Política Social (Ministry of Health and Social Policy), pp 667–722

IMSERSO (2005b) Libro Blanco de Atención a las Personas en Situación de Dependencia en España. Ministerio de Sanidad y Política Social (Ministry of Health and Social Policy), p 670

IMSERSO-CIS (2006) Encuesta de condiciones de vida de las personas mayores 2006

Instituto Nacional de Estadística (2009) Panorámica de la discapacidad en España. Encuesta de Discapacidad, Autonomía Personal y situaciones de Dependencia 2008. Boletín informativo INE 4

Jiménez-Martín S (ed) (2007) Aspectos Económicos de la Dependencia y el cuidado informal en España. Universidad Pompeu Fabra, Barcelona

Keeble L, Loader B (eds) (2001) Community informatics: shaping computer-mediated social networks. Routledge, London

Langa D, Martínez D (2009) Redes Familiares, cuidados y clases sociales en Andalucía In: De la Fuente Y (ed) Situaciones de dependencia y derecho a la autonomía: una aproximación multidisciplinar. Alianza Editorial, Madrid, pp 371–396

Lazarus R, Folkman S (1984) Estrés y Procesos Cognitivos. Springer Publishing Company, Nueva York

Losada A, Montorio I, Izal M, Márquez M (2006) Estudio e intervención sobre el malestar psicológico de los cuidadores de personas con demencia. El papel de los pensamientos disfuncionales. IMSERSO, Madrid

Mata R, García S, Vera P, Romero S (2009) Tecnologías de la Información y de las Comunicaciones para situaciones de dependencia o autonomía reducida. In: De la Fuente Y (ed) Situaciones de dependencia y derecho a la autonomía: una aproximación multidisciplinar. Alianza Editorial, Madrid, pp 179–214

Mestheneous E, Triatafillou J (2005) Supporting Family Carers of Older People in Europe- the Pan-European Background. Supporting Family Carers of Older People in Europe. Empirical Evidence, Policy Trends and Future Perspectives. Available via http://www.uke.de/extern/eurofamcare/documents/nabares/peubare_a4.pdf. Accessed 13 Feb 2013

Ministerio de Industria, Turismo y Comercio (Ministry of Industry, Tourism and Trade) (2010) Plan Avanza 2 Estrategia 2011–2015, Anexos, p 19

Mnich E, Balducci C (2006) Eurofarmcare. Services for supporting. Family carers of older dependent people in Europe: characteristics, coverage and usage. Typology of family care situations. Deliverable no 22

Montorio I, Izal M, López A, Sánchez M (1998) La entrevista de Carga del Cuidador. Utilidad y validez del concepto de carga. Anales de Psicología 14:229–248

OCDE-Organización para la Cooperación y el Desarrollo Económico (2003) Information and communications technologies ICT and economic growth. Evidence from OECD countries, industries and firms. OECD library, Francia

Raya E, Santolaya M (2009) La sociedad de la información y sus aportaciones para el Trabajo Social. Portularia, Revista de Trabajo Social 9:83–92

Read J, Blackburn C (2005) Carers perspectives on the internet: implications for social and health care service provision. Br J Soc Work 35(7):1175–1192

Rogero García J (2010) Los tiempos del cuidado. Impacto de la dependencia de los mayores en la vida cotidiana de sus cuidadores. IMSERSO, Madrid

Schmidt A, Barbabella F, Hoffman F, Lamura G (2011) Analysis and mapping of 52 ICT-based initiatives for family caregivers. Deliverable 2.3-Draft. European Centre for Social Welfare Policy and Research

Chapter 9
Lessons from the Digital Divide

Eduard Aibar

9.1 Introduction

Attention to the digital divide arose at the second half of the 1990s when two different phenomena began to be remarked, highlighted and commented by a wide variety of authors and institutions, through many reports and scholarly publications. On one hand, the uneven diffusion of computer resources and, more specifically, of Internet access, both globally across countries and within nations, and, on the other, the increasing importance of a new and emerging social structure, often named as the *network society* or the *information society*, linked to the development and implementation of ICT, with profound consequences for the economy, culture, politics and social interaction.

Why was the digital divide considered to be an important phenomenon? First of all, because early surveys showed a very specific pattern for most Internet users: a great majority of them were disproportionally wealthy, male, white and, better educated when compared with standard demographic data. Secondly, because the social consequences of the divide were thought to produce severe inequalities in the near future. As many authors put it, the digital divide "is about the gap between individuals and societies *that have the resources* to participate in the information era and those that do not" (Chen and Wellman 2005: 486).

The digital divide has been thought to mean exclusion from the so-called knowledge economy and that holds and bears serious consequences for individuals, social groups and whole nations. The main consequence of not being able to access ICT and the Internet is basically a lower level of participation in the most relevant fields of society. For people, being in the wrong side of the digital divide may entail difficulties in getting jobs, seeking information, accessing public services or participating in political activities. Access to the Internet can also

E. Aibar (✉)
Department of Arts and Humanities, Universitat Oberta de Catalunya, Barcelona, Spain
e-mail: eaibar@uoc.edu

A. López Peláez (ed.), *The Robotics Divide*,
DOI: 10.1007/978-1-4471-5358-0_9, © Springer-Verlag London 2014

facilitate re-defining careers, accessing continuous educations and, in general, encourage personal growth and wellbeing.

In an information society, information is considered to be a *primary good*. That means it is crucial for the survival and well being of individuals and cannot be exchanged for other goods. Information has also been considered an essential source of productivity and power (Castells 2001: 241). All in all, traditional social differences in economic resources, capital and power seem to be actually amplified when access and use of digital technology is added.

Information has also been conceptualized as a *positional good*. That means that different positions in society might entail better conditions for accessing, processing and using information. In a network society, the position social actors have in networks of exchange and information can thus increase or dramatically decrease their relative power and their survival.

9.2 The Digital Divide as a Persistent Reality

The Internet has spread around the globe since 1998 following a raw exponential rate. The total number of Internet users in the world has surged from 900,000 people in 1993, 360 million in 2000, to more than 2.267 million in 2011 (Internet World Stats 2011). However, widespread diffusion does not mean ubiquity.

In more recent times, the digital divide is sometimes presented as something belonging to the past: as if the digital divide was now small and diminishing and thus becoming increasingly irrelevant. In fact, the digital divide seems not to be a hot topic anymore and has slightly lost interest and significance in academic as well as in policy circles (Chen and Wellman 2005: 467). This declining attention is partly caused by the fact that in most developed countries, the spread of Internet access and use has achieved a great majority of the population. The penetration rate in Europe is 61.3 % and in North America has raised until 78.6 %.

However, if we take into account data for all countries and the whole world population, the digital divide has not disappeared at all nor has it diminished to negligible levels—quite on the contrary. Nowadays—December of 2011—in Africa, the penetration rate is 13.5 % and in Asia 26.2 %. The world penetration rate is now about 33 % and that means a vast majority of world population are not Internet users.

When national rates are considered, the differences are even more acute. The digital divide has not certainly vanished. In a study of the role of income inequality in a multivariate cross-national analysis of the digital divide, (Fuchs 2009) comes to the conclusion that "it is unlikely that the digital divide will be closing as long as there is a high degree of global inequality and high degrees of national inequality in many countries" (p. 54). The uneven diffusion of the Internet still persists around the globe—not only between developed and developing countries but even within developed nations. As we will see later in some aspects, it is still growing.

9.3 Digital Divide and Social Inequities

The digital divide, that is, inequalities in access and use of ICT and the Internet, takes place within a much broader spectrum of social inequalities: international and intranational socioeconomic differences, cultural diversity, educational disparities, etc. Much of the academic study of the digital divide has been addressed to measure the influence and correlation between all these sources of social differentiation and those reflected in the digital gap. The literature has identified several elements affecting the digital divide for individuals. The most influential seem to be socioeconomic status, gender, life stage, ethnicity and geographic location (mainly on the rural/urban dichotomy).

Income is the most important singular factor affecting social access to the Internet (Fuchs 2009). Higher-income people are much more likely to be computer and Internet uses than low-income people. Some estimates show that within countries, the inequalities of Internet access are likely to be twice as high as inequalities in income. The level of education has also been identified as another crucial element, since Internet users show a higher level of training through formal educational institutions.

Socioeconomic status is thus the most basic source of inequality for the digital divide—being income and education the most important factors within this category. Internet users are, in general, better educated than non-users and tend to have higher salaries. The socioeconomic influence in the divide seems to be more important for countries with lower rates of Internet penetration: there the differences between users and non-users correlate with greater income and educational inequalities (Chen and Wellman 2005).

Although the influence of income in the digital divide has often been thought to be diminishing because of the declining costs of computers, it still is the most fundamental factor affecting material access since the total cost of computers, Internet access and peripherals remains more or less the same. In 2005, van Dijk (2006: 226) estimated that there was a gap in the most developed countries of about 50 % between the highest and the lowest social strata, since a 90 % diffusion rate was found for the first and only a 40 % one for the late. The figure rises until 90 or 95 in a majority of Third World nations.

The international and global digital divide between developed and developing countries seems to remain substantial. After a multivariate regression analysis on 126 countries in 2005, Fuchs (2009) found that GDP per capita was the most important factor influencing the digital divide. Nevertheless, it was not the only significant factor: social inequalities measured by the Gini coefficient, the level of democracy, and the degree of urbanization were also identified as key factors. The results of this study demonstrate that single models of the digital divide, those reducing causes to single variables, are not accurate enough. Complex models considering the interaction of different socioeconomic, political, cultural, social and technological factors are needed.

But even within developing countries, the divide still seems wide and deep—wide because only a small percentage of the total population are Internet users and deep because the consequences of not being an Internet user may be more important in terms of social and career opportunities (Chen and Wellman 2005: 488). Different studies show that in developing countries, the physical access divide is still widening.

This issue becomes even more apparent when we focus on particular social arenas where the digital divide can be noticeable. For instance, at least for some authors, the digital divide in higher education, rather than losing significance, is today gaining more importance (Selwyn 2010). Considering the issues at stake in the field of education and its wider social repercussions, the consequences of this growing gap may be particularly serious. ICT and the Internet allow students to access a large diversity of educational options and to adapt them to their personal situation—time, place and pace of leaning. As Selwyn puts it, after reviewing a number of empirical studies on the digital divide within higher education: "ICT use continues to be a source of subtle but significant social inequality amongst university students in enduring ways" (p. 39).

9.4 From the Single Divide to a Multifaceted and Complex Divide

From the vast amount of academic literature on the digital divide, two main conclusions can be drawn that may be of great interest when exploring other technology-based social gaps, such as the robotics divide. First of all, the digital divide is, on many grounds, still large and important; the almost exponential growth in the social spread of the Internet in the last two decades has not turned it into a negligible phenomenon. Secondly, from a methodological point of view, it is *multifaceted*. There is not a single digital divide: there are instead many different digital divides.

Until very recently, most studies on the digital divide have been almost exclusively concerned with *access* to the technology—basically to computers or to the Internet. On the one hand, surveys were mostly designed to identify and place people in two broad categories: those with access and those without it. On the second hand, the issue of access was mainly tackled as a simple dichotomy and on a static basis—once access was achieved people were thought to remain there for good (van Dijk 2006: 222).

More recent analyses of the divide have instead come to the conclusion that access is not the only important element. The digital divide is no more taken as a simple binary yes/not question of having access to hardware. Recent research has often remarked that the digital divide in present societies refers to much more than access to certain piece of machinery, whereas a desktop or laptop computer—or even than having basic skills and familiarity with common hardware and software applications. Most of the effective uses of the Internet require not only access to it and some basic computer skills, but also social and cognitive skills. The digital divide is not only a matter of who access the Internet, but of how its use may affect socioeconomic cohesion, inclusion/exclusion, alienation, prosperity or social success.

Usage is in fact the missing term in early approaches to the digital divide. We know now that having access to computers or the Internet is not the same as being able to use them in meaningful or useful ways. As happens with most technologies, one single artefact may be used in a broad spectrum of ways, with a great variety of objectives by different people (Oudshoorn and Pinch 2003). New uses are continuously appearing for old technologies. Not only that, but especially in the field of ICT, we see the emergence of unexpected uses of technology. Some of them are surprisingly different from existing ones, and particularly from those expected, anticipated or recommended by its designers or producers.

9.4.1 Non-Users Matter

This recent shift of attention from plain technology and access, to use and practices of use, has produced a number of important consequences in the conceptualization of the digital divide and the categories associated. One of them is the problematization of non-use and, consequently, of non-users. As happens with many other technologies, non-users have generally been viewed under the assumption that their situation necessarily involves some kind of deprivation or deficiency. In the digital divide arena, those cases in which people with potential access to the Internet voluntarily decide not to be online have hardly been analysed in early studies. We must accept that some people may be not interested at all, while others may feel uneasy when using the Internet—for a number of reasons—or may even express a lack of trust. In fact, a recent survey (Brandtzaeg et al. 2011) shows that 30 % of non-users in five developed countries in Europe do have access to the Internet.

Some taxonomies have been proposed to unveil the complexities of non-use. Wyatt (2003) identifies four different types of non-users. First, what she calls *resistors*—people who have never used the Internet because they never wanted to; second, *rejectors*—those who do not longer use it because they find it uninteresting, boring, risky or expensive; third, the *excluded*—people who have never used the Internet because they could not access it; and fourth, the *expelled*—those who have stopped using it involuntarily because of economic cost or institutional barriers. The widely accepted perspective that universal use has to be the norm for almost every single technology or artefact has prevent many authors from taking non-users as a relevant social actor in explaining uneven social diffusion.

9.4.2 Rethinking Access

In this more recent wave of scholarship on the digital divide, the very concept of access has also been reconceptualized. Though the early studies on the digital divide devoted most of the attention to the technological aspect and then equated technology access with physical access to computers, we do know no, not only that similar levels of access may engage the Internet in radically different ways—equal access does not ensure equal usage-but that different types of access have to be taken into account.

This issue becomes apparent when considering *motivational* access (van Dijk 2006). Before getting physical access to computers and its networks, some kind of wish to be online is needed (even material access itself can be decomposed in physical access to computers and conditional access which refers to different degrees of connection disposal—depending on subscriptions, accounts, broadband, etc.). Being connected is then not only a matter of technological resources availability: as we have already remarked, some people unconnected might have actually refused to be online for different reasons. In addition to those mentioned above, they may see no significant and useful consequence of being online, they may feel they do not have enough time, they may feel computers or the Internet as dangerous, they may think they do not have the necessary skills or operational knowledge, or they may think that the economic cost is too high. Some studies have also identified psychological explanations like technophobia and different kinds of anxiety in front of computers. Some European and American studies between 1999 and 2003 showed that half of the unconnected survey respondents deliberately refused to be connected.

Therefore scholars have increasingly paid attention to more social, psychological and cultural backgrounds surrounding access and use of ICT. The almost exclusive tendency, some years ago, of focussing on the number of people who access or use computers or the Internet and on frequency of use, is no more considered the best way to carry out research on the digital divide. The understanding of access has also moved from a single event of becoming user of a particular technological artefact to a temporal process being influenced by a number of non-technical factors.

Just as material access is preceded by motivational access, it is also succeeded by *skills access* and *usage access* (van Dijk 2006: 224). The increasing analysis of use, beyond access, has brought to the front the issue of skills. Skills can also be decomposed in different types following a certain stage model. Operational skills are needed first when accessing a computer, but then information and strategic skills become crucial for meaningful uses of computers and network resources.

9.4.3 Skills

Skills involved in computer and Internet use are diverse and cannot be reduced to a single set of abilities. A first class of skills encompasses what have been called *instrumental* or *operational* skills. These are the ones needed for operating with hardware and software in a very basic level. Another set of abilities has been named *structural* of *information* skills. Formal information skills involve the knowledge to operate with Internet files, Webs and hyperlinks, whereas informal information skills refer to the ability to find, get, select and assess information in specific Web pages. A final category is that of *strategic* skills: those needed for addressing particular objectives and goals according to the user's expectations, needs and intentions.

Research into information and strategic skills has been much scarcer than that into the traditional digital divide of physical access. And, we must not forget that skills are always a much more difficult issue to research than equipment or technology availability, since users' self-reports of skills, like those provided by surveys, are not often that reliable. Nevertheless, much of the studies and tests that have been carried out show that the skills divide is usually much larger than the physical access divide. They also strikingly prove that, while the access divide is almost closing in most developed nations, the skills gap is growing—particularly when considering information skills (van Dijk 2006; Brandtzæg et al. 2011).

In line with the results of former digital divide studies, the education level shows a strong correlation with the level of skills: users with a high level of traditional literacy do also possess the highest levels of digital information skills. Nevertheless, the concept of *computer literacy* itself has also been called into question by many of these studies on skills. The effective use of ICT and the Internet is no more thought to be based on a relatively simple set of abilities to operate a computer. A *multi-literacies* view tries to emphasize the whole variety of competencies that individuals need to access the digital world and use it in productive and useful way. Carvin (2000) has outlined three basic types of competencies: the *information literacy* that allows users to discern the quality of the content available, the *adaptively literacy* that allows them to develop new skills while using ICTs, and the *occupationally literacy* that allows users to apply these skills in their personal life environment.

Finally, inequality in Internet access and uses does not solely depend on individual skills, capabilities or attitudes. Specific social or institutional contexts may have a crucial role in fostering the access to the Internet. It has been remarked, for instance, that informal training through various ways of social support can provide people the necessary computer and navigation skills to effectively use the Internet.

9.4.4 Usage

Once motivational and material access and skills have been discussed, the final stage, usage, can be treated. We have already pointed out that potential use—availability of computer and Internet resources as well as the skills needed—does not automatically result in actual use. The digital divide tends to appear smaller when actual use is not taken into account. That happens for instance with gender: while the gender gap may look almost inexistent as far as physical access is concerned in many developed countries, when use is considered and deeply analysed, the gap resurfaces again. Generally speaking, once usage comes to the forefront, many traditional social inequalities are translated into the digital divide. So, once more, technology becomes a kind of mirror reflecting—and somehow reinforcing—persisting social differences and inequities (Bijker 2010).

Another aspect of the usage digital divide is the difference between a *consumerist* and passive use and a *productive* or active one. Not all users are equal and similarly creative. This is particularly relevant considering the present spread of Web 2.0 applications and tools, where users are meant to provide most of the content. Though the recent growth in popularity of social networking sites may have partially corrected it, some former studies showed a significant gap between passive and basic uses of the Internet (email, Web browsing, searching) and those more active and creative (contributions to Web sites, blogs, wikis, peer-to-peer networks, etc.) that often involve the collective production and sharing of knowledge and information.

Finally, a certain democratic divide has also been linked to the digital divide when usage is taken into account, since people online have more access to resources in order to engage and participate in public political life. Recent Arab springs, the occupy movement or the Spanish 15 M protests, have highlighted the increasing role of the Internet and social networking in fostering political activism.

This more recent focus on usage has lead many scholars to distinguish and categorize different kinds of Internet users in an approach often called *user typologies*. In a study of Internet users in five European countries (Norway, Sweden, Austria, United Kingdom and Spain), and by means of a cluster analysis on a large and representative sample of 12,666 people, Brandtzæg et al. (2011) identified five types of Internet users.

Non-users are the largest category with 42 % of the sample. People within this cluster do not use the Internet on a regular basis. *Sporadic users* include 18 % of the sample and are formed by people who use the Internet occasionally and infrequently, mainly for e-mail and some other specific tasks. *Entertainment users* group about 10 % of the sample; these are users who show particularly high scores on using Internet radio or TV; downloading games or music; and chatting. *Instrumental users*, about 18 % of the sample, are those who prefer to carry on goal-oriented activities such as searching for information—about goods or services—and use e-commerce, net-banking and travel services. Finally, advanced users, grouping 12 % of the sample, show the highest scores for almost all Internet variables and thus prove to have a very varied and broad Internet behaviour.

This kind of typologies show not only the different ways in which the Internet may be used, but, most notoriously, the inequalities among users when exploiting the benefits of the net. Considering this study analyses data from five well-developed countries with high GPD, it is surprising that 60 % of the sample is found to be non-users or sporadic users. The figure means that a great majority of citizens do not have enough level of usage for a truly effective digital participation. The authors compare this finding with results from a similar study done five years before and discover only a small decline of 2 % for these two groups. Once more, when usage is taken into account, the digital divide seems a resilient phenomenon indeed.

Though the study reflects the common perception that youngsters are more interested and have more opportunities to learn and explore this new technological media and that access is still an important factor, the existence of contextual variations between countries suggest there must be different cultural variables at play. Gender is also found to be another persistent factor in the digital divide. The fact that in countries known to have more gender equality—like Sweden—gender does not prevent users to become advanced users—something that happen in the other four countries—suggests that previous social differences are also easily transferred to cyberspace.

Finally, the study points to another important aspect of the usage divide. As Internet services and applications evolve, it is very likely that the levels of usage of the different user typologies will increase. Nevertheless, since people who are more active and advanced users will achieve new competencies and skills faster, the divide in terms of the different user categories will most likely grow. New kinds of inequality in the use of the Internet are thus expected in the near future.

9.5 Wrong Assumptions for Tackling the Digital Divide

The maturity reached in the analysis of the digital divide has not only provided more data and insights on new aspects of the divide, but an awareness of the conceptual and methodological flaws of earlier approaches. We have already pointed out some specific ones, and now, we will discuss others with a more basic character.

9.5.1 Internet Access as Multimodal Phenomenon

Access, for example, has been rethought in a number of ways—from a single event, to a process and from a yes/not question, to a gradual attribute. However, the evolution of technology has also put into question another aspect of access.

Although fifteen or ten years ago accessing the Internet was mainly possible through personal computers, nowadays different pieces of equipment and artefacts are available for Internet connection: not only personal computers or laptops, but also mobile and smart phones, tablets, e-books, TV's, game consoles, etc. Any meaningful approach to the digital divide has to acknowledge the present convergence of new media platforms and technologies that allow people to go online.

Computer technology, and particularly Internet access and use, is nowadays a *multi-modal* phenomenon. This is not only a purely technological matter: access and use of the Internet can be greatly affected by the artefact used for establishing the connection. An Internet search through a smartphone is not the same than through a desktop computer. Screen size, software available in the device, speed, cost, size of messages, etc., provide some restrictions and limits for Internet navigation. Depending on the particular artefact being used for connecting to the Internet, there is a wide differing range of technical and social qualities available.

9.5.2 Domestic Divides are also Important

Until very recently, the digital divide was mainly explored following the boundaries of national states. In the last years, many studies have also explored more detailed inequalities within nations. Many authors have argued for a redefinition of the digital divide concept in order to take into account differences between individuals—for instance, the inequalities of Internet use within rather than between nations. The focus has also shifted from developed countries—where the Internet was first diffused—to the developing word.

9.5.3 Taking the Divide Metaphor too Literally

Metaphors are common in many areas of scientific thought. When considering the relationship between technical innovations and social change, some spatial and mechanical metaphors have been traditionally used. We frequently talk of the social *impacts* of a technology—though technology does not come out of the blue, from a non-social medium, and though social impacts of technology do not happen at a single point of time. When taking too literally, metaphors can contribute to confuse the phenomenon at stake. This has often happened in digital divide analyses. The expression "digital divide" may suggest, and indeed, it has frequently suggested, a too dichotomized and sharp distinction between two clearly separated social groups. It has also suggested that the divide is about absolute inequalities between those included and those excluded, whereas most of the inequalities identified by recent studies show a relative and dynamic nature (Brandtzæg et al. 2011).

Empirical research shows that these are oversimplifications of a much more complex reality. The sharp distinction between users and non-users hides significant and more subtle types of relationship to any particular technology. Intermittent users, for instance, are those who become non-users for extended periods of time. In fact, we must not assume that everyone having a piece of machinery is actually using it: possession of the artefact does not equate automatically to use. Many people make a very rare use of computers at home and some do not use them at all. Frequency of use may actually resurface divides that are masked when only access is considered.

9.5.4 Technological Determinism

Technological determinism has been the most implicit theoretical assumption behind many analyses of the digital divide—in fact, technological determinism is still the most influential and popular view of the relationship between technological innovation and social change. It encompasses two different elements: (a) technology develops autonomously following a sort of intrinsic logic and (b) technological development is the most important singular element determining social change (Bijker 2010: 71).

One of the traces of technological determinism can be detected in the almost exclusive emphasis early studies made on access to technology (computers and connection to the Internet, mainly). This has been considered for long the most critical aspect of the divide; this tendency has minimized the other non-physical aspects of the divide that we have already commented. Furthermore, given the multifaceted nature of the digital divide, simple policy measures for narrowing it may be less successful than expected. Since access does not equate with usage, solving the divide is not only a matter of providing computers and Internet connections. The promotion of an effective use of the Internet is certainly more complicated and cannot be reduced to simple policies for fostering access.

Another related technological deterministic idea has been the often implicit consideration that access to computers, per se, would trigger important social and economic benefits in a highly mechanistic and causal connection. This scheme underlies a vast majority of the digital divide research. Such an emphasis on the role of technology in overcoming social differences and inequities tends to obscure two recent and empirically based views in the sociology of technology: on the one hand, the fact that technical artefacts are usually shaped by particular values and visions—sometimes well rooted in the status quo, and, on the other hand, the fact that new technologies may also contribute to reinforcing and consolidating existing social differences (Aibar 2010).

9.5.5 The Linear Model of Technological Development

Many digital divide studies have also shared an understanding of ICTs evolution and diffusion as a *linear* process where these technologies increasingly spread through the whole society. Sometimes, digital divide researchers have even resorted to diffusion of innovation theory and its most fashionable version, the S-curve—an incredibly popular perspective in business and marketing literature—as a simple model for the social adoption and acceptance of ICT. Under this deterministic view, technology diffusion is implicitly assumed to follow an autonomous path of necessary and mechanistically joined steps, so further research into social shaping forces or actors is not considered necessary or very relevant.

The social actors or social aspects involved are instead considered sometimes only as *obstacles* to the autonomous and unidirectional path of technological innovation, which seems to be powered by an internal *momentum*. Therefore, some authors talk about the cultural or social "barriers" to ICT or the Internet, for instance, and others resort very easily to different kinds of social or human "resistances" against technological innovations.

9.5.6 Methodological Flaws

Van Dijk (2006) points to several theoretical and methodological flaws of digital divide research and analysis. First of all is lack of theory. Much of the research has got a very descriptive nature and the analysis of deeper social, cultural or psychological causes influencing the divide has been very much neglected. The lack of qualitative research has somehow obscured the actual mechanisms in the process of appropriation and domestication of ICT-related technologies. And, most quantitative studies, though presenting a great deal of correlations, have not included longitudinal data, particularly necessary in such a changing technological landscape.

Another problem is the lack of interdisciplinary research. The preponderance of sociological and economic research has underestimated psychological, educational and cultural factors affecting the digital divide and, more specifically, the usage dimension.

9.6 Conclusions: Lessons for Analysing the Robotics Divide

The lessons we can learn from the last 15 years of digital divide research and analyses are particularly relevant for present and future efforts to address the robotics divide. Not only because robotics and ICT are two areas of technology deeply interconnected for obvious technical reasons, but because the social diffusion of robots is increasingly foreseen in the light of what we know about the diffusion of ICT and the Internet. Bill Gates has been quoted to observe that "the emergence of the robotics industry [...] is developing in much the same way that the computer business did 30 years ago" (quoted in Lin et al. 2011: 942).

It is very likely that many experts agree with such a perspective—considering it comes from one of the key figures in the computer business. But what is more worrying is that many of these experts seem to be falling again in some of the typical mistakes that we observed in the analysis of the digital divide. Not only exponential progress, following dubious laws of technological development such as Moore's law, is forecasted but ubiquitous diffusion of robots throughout society is also expected.

Most of the flaws in digital divide research we have identified in this chapter rest on very well-known views of the relationship between technology and society, that is, between technological innovation and social change. Some of these old views, I have argued, are based on assumptions that are not necessarily consistent with what we already knew about technology and society and the many ways they interact.

Science and technology studies (STS) have been critically analysing the interaction between science, technology and society for the last four decades and have developed a deep and very rich picture of its intricacies (Hackett et al. 2008). I have used their approach in order to identify some of these persistent misconceptions and, thus, show some opportunities for improving future research in the robotics divide.

A first general lesson we may draw from STS is that technological innovation is not a single point-in-time event but a process occurring over time and subject to many heterogeneous forces. Technologies are not "ready made" at one point after their design and development phase, and then spread throughout society. Implementation and use often produce changes in design (Bijker 2010).

In the last decade, there has been an increasing amount of scholarship devoted to the understanding of user–technology relations—this has represented a remarkable shift from the older and more usual study of designers by most social analysts of technology (Oudshoorn and Pinch 2003). This change in orientation has also occurred in the social study of ICT. But although there are many studies of users and uses of ICT in very different areas, there are still some insights into this broader literature on users and technological artefacts that could be particularly useful for tacking other technological divides.

First of all, users should be understood as active and not passive participants in the evolution of technology. They are not simple consumers, but active agents in the *domestication* and adaptation of artefacts to their own objectives and interests. We should not forget that the very origin and evolution of the Internet shows this remarkable blurring of the distinction between users and producers (Abbate 1999).

Secondly, social scientists should place more emphasis on the disaggregation needed to understand the many possible uses of any technology. Another important lesson from STS in this area is that there is never a "correct" use of a technology: there are only intended, recommended, expected or dominant uses. Use is never deduced from the technology itself and though designers or producers invest a lot of time and resources to discipline their future users, it is always possible that they end up with totally new and surprising uses.

Another important lesson to be learned from STS concerns the alleged revolutionary power that we tend to confer to technology—whether ICT or robotics. The way we academics tend to phrase our research questions is often too grandiloquent—it seems that we have been infected by the same virus that affects enthusiastic journalism, supply-side marketing and oversimplified policy visions, maybe because academic social science is often in dialogue with them. Whenever we envisage changes (linked to technology), it seems they have to be big, revolutionary and dramatic. And this kind of research megalomania affects not only the deepness of those changes but their scope. Things are expected to change a lot and worldwide.

A useful recommendation to avoid this kind of pitfall should be again *disaggregation*. I think we need to disaggregate society, users and even technologies much more, otherwise our conclusions lose relevance and soundness. Whenever we talk about important changes or impacts, we have to specify how important they are, in what particular circumstances and for whom—this applies to both ICT and robotics (López-Peláez and Kyryakou 2008).

In that context, the *performative* character of technology narratives should also not be forgotten. Not only does technology have social effects, but so also do discourses about them. In the field of e-Government, for instance, the aggressive and deterministic views and stories produced by consulting, software and hardware companies have had a very deep influence in the way ICT and the Internet have been used in the last decade by many governments (Waksberg and Aibar 2007).

Another remarkable consequence of the deterministic "impacts" frame is the treatment of technology social effects as universal, predictable and unidirectional. In most cases, this is inaccurate. A large proportion of STS empirically based case studies have been, in fact, devoted to demonstrating that the uses and effects of technologies depend decisively on local social contexts. ICTs alleged effects cannot be seen as independent of the social environment where they have actually been designed and created. We need more informed studies not only on technology effects on society, but on the way, technologies themselves are actually designed, developed, tested and thus shaped along those processes. We need also to bear in mind that innovation is not only a scalar magnitude but a vector, that is, something that has got another property worth of mention: direction.

Finally, maybe, the most important contribution of STS has been to prove that the link between technology and society is always twofold. Technology impacts society but society, in its turn, shapes technology. That simple thesis has got important implications when analysing technological divides and their relationship with social inequalities.

It is not only that different social inequalities greatly explain technological divides, but that those technological divides, themselves, may have a deep impact in the continuation or even the deepening of those social inequalities. Though ICTs have often been considered a potential source for change in many social arenas, it has also been proved that they can reinforce existing organizational structures or power relations in other contexts.

References

Abbate J (1999) Inventing the Internet. MIT Press, Cambridge (MA)

Aibar E (2010) A critical analysis of information society conceptualizations from an STS point of view. Cogn Commun Co-oper 8(2):177–182

Bijker WE (2010) How is technology made?—That is the question! Camb J Econ 34:63–76

Brandtzæg PB, Heim J, Karahasanović A (2011) Understanding the new digital divide—A tipology of Internet users in Europe. Int J Hum Comput Stud 69:123–138

Carvin A (2000) More than just access. Educause, Nov/Dec 38–47

Castells M (2001) The Internet galaxy: reflections on the Internet, business, and society. Oxford University Press, Oxford

Chen W, Wellman B (2005) Charting digital divides: Comparing socioeconomic, gender, life stage, and rural-urban internet access and use in five countries. In: Dutton WH, Kahin B, O'Callaghan R, Wyckoff AW (eds) Transforming enterprise: The economic and social implications of information technology. MIT Press, Cambridge (MA), pp 467–497

Fuchs C (2009) The role of income inequality in a multivariate cross-national analysis of the digital divide. Soc Sci Comput Rev 1(27):41–58

Hackett EJ, Amsterdamska O, Lynch M, Wajcman J (2008) The handbook of science and technology studies. MIT Press, Cambridge (MA)

Internet World Stats (2011) World Internet usage and population statistics. http://www.internetworldstats.com/stats.htm. Accessed 19 Jun 2012

Lin P, Abney K, Bekey G (2011) Robot ethics: mapping the issues for a mechanized world. Artif Intell 175:942–949

López-Peláez A, Kyriakou D (2008) Robots, genes and bytes: technology development and social changes towards the year 2020. Technol Forecast Soc Chang 75:1176–1201

Oudshoorn N, Pinch T (2003) Introduction: how users and non-users matter. In: Oudshoorn N, Pinch T (eds) How users matter: the co-construction of users and technology. MIT Press, Cambridge (MA), pp 1–25

Selwyn N (2010) Degrees of digital division: reconsidering digital inequalities and contemporary higher education. Revista de Universidad y Sociedad del Conocimiento 1(7):33–41

Van Dijk J (2006) Digital divide research, achievements and shortcomings. Poetics 34:221–235

Waksberg A, Aibar E (2007) Towards a network government? A critical analysis of current assessment methods for E-government. In: Wimmer M, Scholl HJ, Grönlund Å (eds) Electronic government. Lecture notes in computer science, vol 4656, Springer Heidelber, pp 330–341

Wyatt S (2003) Non-users also matter: The construction of users and non-users of the Internet. In: Oudshoorn N, Pinch T (eds) How users matter: the co-construction of users and technology. MIT Press: Cambridge (MA), pp 67–80

Chapter 10
Inequalities in the Information and Knowledge Society: From the Digital Divide to Digital Inequality

Cristóbal Torres-Albero, José Manuel Robles and Stefano De Marco

10.1 Introduction[1]

In this chapter, we exemplify the set of problems addressed in this book regarding the inequalities generated by technology and, more specifically, by robotics, through the study of the social consequences that the technological vehicle, that is, the Internet, has brought about in advanced contemporary societies in the course of the last decade, taking Spain as a case study. This provides us with a comparative approach to the social inequality outcomes that the Internet, as a fully implemented technological tool, has already produced.

Indeed, with the emergence of the Information and Knowledge Society, two notions have captured the attention of social scientists and persons responsible for public policy linked to the development of this new type of society. These are the concepts of Digital Divide and Digital Inequality. The first of these refers to the differences between people who access and people who do not access the Internet (Van Dijk 2006). On its part, the concept of Digital Inequality was coined to analyse the inequalities arising between Internet users as a result of the different uses made of this medium.

---

[1] This chapter is part of the work undertaken under research project CSO 2009-13424 of the Spanish Ministry of Economy and Competitiveness.

C. Torres-Albero (✉)
Department of Sociology, Faculty of Economics, Universidad Autónoma de Madrid (UAM), Madrid, Spain
e-mail: cristobal.torres@uam.es

J. M. Robles · S. De Marco
Department of Sociology III, Faculty of Economics, Universidad Complutense de Madrid (UCM), Madrid, Spain
e-mail: jmrobles@ccee.ucm.es

S. De Marco
e-mail: s.demarco@cps.ucm.es

A. López Peláez (ed.), *The Robotics Divide*,
DOI: 10.1007/978-1-4471-5358-0_10, © Springer-Verlag London 2014

In the seminal works, the concept of Digital Divide was applied explicitly to the differences between certain social groups and certain others, depending on whether or not they had Internet access (Horrman and Novak 1998; Strover 1999). Initially, the concept of "access" was used literally, and therefore, researchers focused on the description of social groups, geographical areas, or countries with greater or lesser Internet access infrastructure (Walsh 2000; Attewell 2001). However, it was soon clear that Internet access did not guarantee Internet use (DiMaggio et al. 2001). Likewise, it was shown that people who had the motivation to use the Internet, even though they did not have access in their homes, workplace, or immediate environment, would seek for alternatives that allowed them to connect (DiMaggio et al. 2001). Therefore, the idea of the Digital Divide as the gap between those with or without access soon lost ground and gave way to the idea of "use". Thus, the Digital Divide is currently understood as the difference between people or social groups who use the Internet and those who do not (DiMaggio and Hargittai 2001).

However, this dichotomous perspective (access versus no access; use versus no use) of the social risks associated with the development of the Internet has also been called into question for several reasons. The main argument is that digital inclusion is an attainable goal. According to this thesis, the inequalities between social groups that use and those that do not use the Internet are reducing, and therefore, we could start envisaging a scenario of universal digital inclusion (NTIA 2000). From this point of view, the Digital Divide is understood as a form of circumstantial inequality, arising as a result of the barriers for technologies to spread among less privileged social groups or those less inclined to incorporating them. Public and private strategies to reduce the Digital Divide, as well as the very social dynamics of "contagion" of Internet habits, have led to researchers putting into perspective the importance of this form of social inequality.

In recent years, there has been a move to start referring to the concept of *Digital Inequality* (van Deursen and van Dijk 2010). This notion involves several premises. The first is the argument already mentioned regarding the realistic prospects for digital inclusion. In other words, we are starting to move towards a scenario where the object of study are citizens who use the Internet. The second of these premises is that not all the uses of this medium are the same. Thus, whereas some favour social capital, professional career progress or the enjoyment of better and cheaper goods and services, others are simply online reproductions of activities that can be done off-line or that, despite being a new type of activity, do not contribute any added social, political or economic value. Thus, Digital Inequality is understood as the differences existing between certain Internet users and others, depending on their capability of obtaining advantages and benefits from the use of this technology (DiMaggio et al. 2004). These specialists put forward the hypothesis that, unlike the Digital Divide, this form of inequality, far from reducing, is actually increasing or, at least remaining steady, between socially better-positioned groups and socially worse-positioned groups.

In this chapter, we outline a debate regarding these two concepts related to the unequal development of the Internet from an empirical point of view, taking Spain as a case study. Our goal is to verify, for the case of Spain, to what extent the differences between the different social groups in terms of Internet use are reducing or not, as well as to ascertain whether Digital Inequality is a growing form of social inequality. To meet this goal, we performed a time series analysis from 2004 to 2011, with the data provided by the Spanish National Statistics Institute (INE). We used the ANOVA method because it allowed us to measure, unlike most methods, the statistic significance of the inter-annual differences of both the Digital Divide and Digital Inequality, as well as to estimate when and between what groups it arises. This analysis was applied to the general evolution of Internet use and, at the same time, to the evolution of one specific type of Internet use, namely e-shopping which, according to the literature, represents a case of Internet use that generates economic advantages.

This chapter continues with a second section where we provide a review of the literature on the concepts of Digital Divide and Digital Inequality in order to build our analysis upon solid theoretical foundations. To do so, we start with two recurring concepts in the literature on the development of the Internet: *normalisation* and *stratification*. Both these concepts have been coined to describe Internet development and penetration patterns in a given society and, therefore, offer different perspectives on what effects an uneven distribution of this technology among the population would have. Thus, these concepts have been used mainly to analyse the social structuration of Internet penetration within the framework of the Digital Divide. In our case, they are also applied to the development of electronic commerce, susceptible of generating Digital Inequality. With both notions, and with the data from our empirical analyses, we draw our conclusions regarding the development of these two types of inequality in the Information and Knowledge Society, as well as of the social scenario they generate.

10.2 From the Digital Divide to Digital Inequality: Normalisation and Stratification

In this section, we provide a brief reconstruction of the evolution of the concepts of Digital Divide and Digital Inequality. To do so, we refer to the contributions of the main specialists on the subject and we summarise the criticisms made to each of the concepts. In addition, we apply the concepts of *normalisation and stratification* (Norris 2001) to the development of the Internet and the analysis of the Digital Divide and Digital Inequality.

According to the concept of *normalisation*, the differences between social groups only increase in the initial phases of the development of the Internet, gradually disappearing as the groups most inclined to use this technology reach saturation levels, understanding as such percentage rates where penetration of Internet use starts to stabilise or at least slow down in terms of inter-annual growth. On the other hand, *stratification* assumes that each social group starts with different Internet use penetration levels, and more importantly, it also considers that they have different saturation points. Thus, it is understood that where certain social groups will reach saturation levels close to one hundred percentage, others will stabilise at lower percentages. From this point of view, then, inequalities in Internet use will tend to reproduce the structural inequalities of a given community. These two concepts allow us to enter the debate as to which type of inequalities appears in the Information and Knowledge Society and what their effects might be.

10.2.1 Definition and Development of the Concept of Digital Divide

As mentioned above, the concept of Digital Divide refers to "the gap between those who do and those who do not have access to new forms of Information Technology" (Van Dijk 2006, p. 221) and started to be used in the 1990s to describe the empirical evidence of the uneven penetration of the Internet in US homes. The first time the term appeared in writing was, according to Gunkel (2003), in the study carried out by the US National Telecommunications and Information Administration (NTIA 1999).

Already then, a significant number of specialists (Horrman and Novak 1998; Strover 1999; Walsh 2000; Attewell 2001) pointed out that the unequal access to the Internet was strongly influenced by geographical variables such as size of the city or geographical region of residence. Likewise, they showed that the best positioned social groups (young people, men, people with higher levels of education or belonging to certain racial groups) had higher Internet access rates than people belonging to less privileged population groups (the elderly, people with lower levels of education, women or those with fewer economic resources). This fact led to the study of the Digital Divide becoming a meeting point for social scientists interested in warning about the potential risks associated with this inequality of access to the Internet.

However, in academic terms, the term Digital Divide is somewhat ambiguous. Van Dijk (2006) points out some epistemological aspects of this concept that generate a biased perspective of the type of inequalities they try to describe. Thus, he holds, among others issues, that this concept suggests an overly simplistic division between two groups of population—people with access and people without access—which is, in addition, too static and, it would appear, very difficult to maintain. It is, from this point of view, a deterministic perspective according to which the problem of digital inequalities results from having or not having access to the Internet. In other words, the provision of this resource determines the solution to this form of inequality.

One of the main obstacles for this concept to become a valid operative resource for the study of digital inequalities is the assumption that access implies use. Soon, many studies showed that citizens with access to the Internet—whether at their homes or at public centres equipped for this purpose—did not necessarily make use of this technology. Around 2003, specialists had already turned their attention to the reasons why certain people or social groups did not make use of the Internet despite having access to it. This new perspective of the Digital Divide made it evident that the differences in Internet use are determined by social variables, whether they are race based (Hoffman et al. 2001), gender based (Bimber 2000; Cooper and Weaver 2003), or education based (Bonfadelli 2002), as well as by another set of variables related to the ability to use the Internet (DiMaggio et al. 2004; van Deursen and van Dijk 2009).

Likewise, it was observed that there are certain individual motivations that affect the inclination to use the Internet. Along these lines, one of the most productive theoretical frameworks is that known as TAM or technological acceptance model (Torres-Albero et al. 2011). From this point of view, the use of the Internet is determined by a complex set of factors among which Perceived Usefulness of Technology and Perceived Usefulness of the Medium are especially relevant. Along these same lines, there has been research carried out into how other psychological and mental factors influence Internet use. This is the case of "computer anxiety" or technophobia (Rockwell and Singlenton 2002).

However, despite the variety of factors put forward to explain the Digital Divide, a significant number of specialists agree that it is a form of inequality that is set, if not to disappear, to at least reduce significantly. In recent years, empirical studies have justified the thesis that both public and private measures for the development of Internet use, and the very processes of social interaction, are leading to an increase in the level of Internet use penetration among those social groups who were initially more reluctant to incorporate it. In this regard, we find quantitative studies (Eurobarometer 2002–2011) that show a process of convergence between people of different sexes, level of education and economic status. Qualitative studies also show how relationships between parents and children turn the latter into veritable educators and promoters of the use of the Internet (Rojas et al. 2004). Indeed, the youngest social groups, more inclined to using ICTs, generate contagious social dynamics in that they perform a role as initiators in the use of these technologies for adults and elderly people. Thus, these inter-generational relationship dynamics within the family are one of the ways by which the resistance of certain groups to using ICTs is cracked.

As a result, from a theoretical point of view, the idea of social inclusion has become a commonplace for many specialists who defend the thesis that the generational renewal and the social development dynamics of technologies shall eventually make the use or non-use of the Internet a relatively innocuous factor for social inequalities.

10.2.2 The New Idea of Inequality: The Concept of Digital Inequality

Especially since the start of this century, this field of study has shifted its attention towards a perspective of analysis that is more complex in terms of the relationship between Internet and inequality. One of the most interesting efforts has focused on analysing to what extent certain uses of the Internet generate competitive advantages for its users (Van Dijk 2005). There have been studies on issues such as how the Internet allows citizens to express their demands and interests in an easy and efficient way, how the Internet is a key factor for enjoying better goods and services, and how the use of this medium allows users to access competitive resources. These types of use of the Internet have been termed *Beneficial and Advanced Uses of the Internet* (BAUI). From this point of view, Digital Inequality would be the result of the difference between citizens who make use of this type of services and tools of the Internet and citizens who do not have the resources to make use of them.

There are different taxonomies that attempt to order the elements that potentially affect the ability of citizens to use BAUI (DiMaggio and Hargittai 2001; Van Dijk 2006). One of the first in the field proposes to classify the dimensions of Digital Inequality into four categories: technical resources; autonomy; social and institutional context; and digital skills and the purpose of use of the Internet (DiMaggio and Hargittai 2001).

These two latter authors consider that the limitations in hardware, software and type of Internet connection become an important barrier for using this type of technology with no limitations. Therefore, the suggestion is that technological equipment is a fundamental factor for understanding what services a citizen uses and, thus, what benefits they can obtain from the use of the Internet. Secondly, and along the same lines, the place from where Internet is used, as well as the control over which pages can be used, has been considered fundamental factors for analysing the types of use of the Internet citizens make. Using the Internet from public places such as Internet cafes, computer centres or from the workplace may imply a loss of autonomy of the user that affects the type of activities carried out online. Likewise, the use of the Internet in public places implies, in many cases, being exposed to restrictions regarding the Internet sites or services that can be used. Equipment limitations and autonomy of use are material restrictions that imply significant limitations in accessing the potential benefits of using the Internet.

Thirdly, social and institutional contexts have been deemed determining factors of the type of use made of the Internet. Citizens who live in a technologically stimulating environment are faster to develop an inclination to use the Internet, as well as greater skills to get more out of this tool. As mentioned above, for adults who have young people in their place of residence, the likelihood of using the Internet more frequently and for more diverse purposes is higher than for people who live in contexts that are technologically less stimulating (Rojas et al. 2004). Likewise, in a favourable institutional context where there are public strategies for the technological education of citizens, it also becomes a key factor for improving citizens' skills to obtain benefits from the use of the Internet.

Lastly, one of the factors that have most caught the attention of Digital Inequality specialists refers to citizens' type and level of skills in using the Internet. It is deemed that the greater and more varied the knowledge and skills for using the Internet are, the more likely they are to access potentially beneficial services. The literature (van Deursen and van Dijk 2009) has generated two broad categories for analysing digital skills. On the one hand, what is known as Internet Expertise and, on the other hand, Internet Proficiency. The first of these measures the degree of integration of the Internet in a user's daily life. However, it is deemed that through the analysis of the time a citizen has been using the Internet, the variety of places from which they connect to the Internet or the frequency with which they connect to the Internet, it is possible to measure their Internet use skills indirectly. On its part, the variety of uses an Internet user makes of the Internet is also an indirect indicator of the Internet user's technological skills. Even though the literature has generated techniques that attempt to measure Internet use skills directly (Van Dijk 2006; Hargittai 2010), these concepts of Internet Proficiency and Internet Expertise are, given the complexity of the methods of measurement, the most common in the literature for evaluating digital skills.

The literature shows how each and everyone of these dimensions is marked by socio-demographic variables such as age, level of education, sex or social class. Thus, we have empirical studies (Robles et al. 2010) that allow us to verify the social character of the dimensions that explain Digital Inequality. It is here where the concept of Digital Inequality takes on its real significance. As shown by the mentioned studies, socio-demographic variables are, in addition to the four dimensions outlined in this section, key factors for explaining the type of use citizens make of the Internet. We find that very often the Beneficial and Advanced Uses of the Internet are performed most frequently by citizens with greater material and educational resources. That is, the BAUI provide potential advantages to citizens who were already privileged beforehand. Thus, the most relevant proposal of Digital Inequality research is that BAUI reinforce and increase existing social inequalities in a given society in that they provide advantages to citizens who enjoy a better social position. The corollary to this argument is that, according to the empirical studies carried out (DiMaggio et al. 2004), this type of inequality, far from reducing, is actually increasing in developed countries.

10.2.3 The Structure of Digital Inequality: Normalisation Versus Stratification?

The first studies on the Internet focused on the idea of the social impact of technologies (Levy 2002). From this point of view, the spread of Internet penetration would have a significant effect on society given that it would change fundamental social patterns and behaviours. Under this same view, the outlook regarding these effects was optimistic, as was the forecast of the way in which they would occur. It was deemed, for instance, that the Internet would allow a revitalisation of citizens' civic engagement brought about by a reduction in the costs associated with political participation (Hague and Brian 1999). Likewise, it was speculated with the idea that the Internet would make it possible to generate a scenario where information and knowledge would become far more accessible and democratic resources (Negroponte 1996). However, this perspective soon clashed with a reality where the Digital Divide showed an uneven distribution of Internet access and use and, therefore, a far less positive scenario than that originally considered (DiMaggio et al. 2001).

Despite the empirical evidence on the Digital Divide, public and private policies for the development of the Information and Knowledge Society have been based on the idea that the Internet should become a vehicle to strengthen citizenship and the political, economic and social possibilities of citizens (NTIA 2002). Within this context, the idea of *normalisation* arose to describe a process of evolution of Internet penetration where the initial differences between groups that are more and groups that are less permeable to the use of this medium tend to decrease as a result of public and private measures to reduce the Digital Divide (Norris 2001). This is again an optimistic outlook on how the use of the Internet would evolve, as well as with regard to the institutional ability to extend this behaviour among citizens. This is also a deterministic perspective to the extent that it deems that the provision of access and the enablement of the use of the Internet will put an end to technological inequalities.

From our point of view, this perspective is based on an idea of formal equality of opportunities (Nozick 1991). This idea suggests that all citizens, regardless of their social condition, should have, at least to start with, the same opportunities. The idea is to guarantee a scenario where all citizens are in a position to use and do actually use the Internet. Inequalities, in the event that they did exist, would be the result of the personal development of citizens, of their interests, of motivations and of their ability to obtain more or less benefits from the use of this medium. In short, this perspective puts forward a proposal as to how inequalities should be managed politically that is focused on guaranteeing an equal starting point and letting individual capabilities position citizens socially.

In opposition to the thesis of normalisation, the perspective of *stratification* starts from a somewhat less optimistic point of view. It assumes that different social groups have different maximum levels of Internet use penetration. These maximum levels may be defined by the very characteristics of the social group, or by the value and subjective use, this group gives the Internet (Norris 2001). Some specialists have also warned of the influence the characteristics of the medium itself has on the perception of the use of the Internet (Robles et al. 2010). It is thus considered that the Internet adjusts better to the interests and expectations of certain social groups, whereas it is removed from the vital goals of citizens pertaining to other social groups.

From this point of view, it is not possible to reach universal digital inclusion but, at the most, to increase the percentage of Internet users among social groups less permeable to the use of this tool. This would imply, in any case, that the development of the Information and Knowledge Society will reflect, despite institutional efforts, the structural inequalities existing in a given community.

10.3 Methodology

Our first empirical goal is to verify whether in recent years there has been a reduction in the differences existing in Spanish society in terms of the Digital Divide and an increase in Digital Inequality for a specific case. As explained above, to do so, we observed the evolution of the penetration of Internet access and use of e-shopping among the Spanish population from the years 2004 to 2011. As pointed out by several authors (Wang et al. 2002; Grewal and Levy 2004), e-shopping is an Internet use that generates substantial economic advantages for its users and improves their opportunities of obtaining better services. In this regard, e-shopping is recognised in the literature as a BAUI.

Our second goal is to find out whether, among the patterns of evolution of Internet access and e-shopping percentages, it is possible to see differences between different segments of the Spanish population. More specifically, the goal is to find out whether the population segments described as more advanced by authors of the Digital Divide and Digital Inequality have or do not have greater percentages and sharper rates of growth. To do so, we have broken down the two time series according to socio-economic and demographic variables deemed relevant in the literature: age, level of education and employment status.[2]

10.3.1 Technical Details of Survey

In order to meet our empirical research goals, we have used the data of the Spanish National Statistics Institute (INE) (2011) obtained through the "Survey on Equipment and Use of Information and Communication Technologies in Households". These surveys gather information regarding Internet access and the

---

[2] Variables included in the Spanish National Statistics Institute (INE) have been used. The three are ordinal and the categories that make up each of them are:

- Employment status: Active employed population, Active unemployed population, Students, Home-makers and Pensioners.
- Age: 16–24, 25–34, 35–44, 45–54, 55–64 and 65–74 years.
- Level of education: No formal education, Primary Education, First level of Secondary Education, Second Level of Secondary Education, Higher Professional Training, Higher Education.

different uses of the Internet by Spanish citizens, including e-shopping, and they cover a time period from 2004 to 2011. For this analysis, we have used the variables "Internet access" and "Purchase of products or services through the Internet". Both variables are dichotomous and both measure the mentioned behaviours in the three months prior to the survey.

The surveys of the different years have the same methods and sampling criteria, the same formulation of the questions regarding Internet access and e-shopping and, lastly, the same reply categories. Thus, the data are respectful with the methodological requirements to be followed for construction of time series.

The eight samples gathered refer to the Spanish population of both sexes, of an age range of between 16 and 74 and living in family homes within the Spanish territory. Only one person per home was surveyed, previously selected randomly by computer. The sample design was carried out through a three-stage sampling procedure stratified by the units of the first stage. These units coincided with the census sections. The units of the second stage are main family homes. In the third stage, a person of over 16 years of age was selected in each household. The stratification criterion used was the size of municipality the section belongs to.

In addition, for each autonomous community (the Spanish regional administration political unit), an independent sample was designed to represent them. The sample was distributed across the autonomous communities using a compromise allocation uniform to proportional to the size of the community.

For homes for which no telephone contact was available, personal interviews were carried out, following the CAPI methodology. For homes for which there was a telephone contact available, CATI interviews were carried out. Both surveys gathered the same data and used the same variables. Figure 10.1 shows the total number of subjects of the sample, stating whether they were surveyed, following the CAPI or the CATI methodology.

It is important to point out that although the sample sizes are not identical, the homogeneity of the sampling method across the different years of data collection justifies the use of these same data to construct a time series. Lastly, it should be

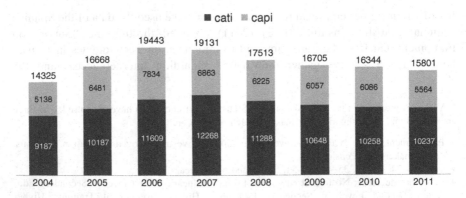

Fig. 10.1 Total sample sizes according to information collection methodology. *Source* INE

noted that weighting factors were used to consider the subjects and not the homes. To do so, an elevation factor was applied, calculated by the National Statistics Institute.

10.3.2 Methodology Used

For the analysis of low-frequency time series, such as those presented here, it is not possible to implement prediction models such as, for instance, the ARIMA model. However, it is possible to draw conclusions regarding the evolution of data from the observation of the trends of the series themselves. Together with these observations, ANOVA tests were performed, in order to point to those differences in percentages, either between years or between the different categories of the control variables, which are significant.

To study the evolution of Internet access penetration and e-shopping rates, firstly, two ANOVA tests were implemented. One takes Internet access as an independent variable, and the other takes e-shopping. In both cases, the survey year was used as a grouping factor. The dependent variables are dichotomous variables, so rather than comparing averages, the analysis has compared percentages. Thus, we have observed the differences in percentages of Internet access and e-shopping between different years, underlining those differences which, from year to year, have turned out to be significant.

Secondly, the ANOVA tests were repeated, also including the age, level of education and employment status variables as grouping variables. The dependent variables were been modified. Thus, different analyses were implemented, one per survey year, to find out whether in the same year and across different categories of each of the three control variables, there had been significant differences in the Internet access and e-shopping percentages. In addition, in order to guarantee the comparison between all the percentages, subsequent ANOVA tests were implemented across the years, selecting, for each analysis, only those subjects that responded to a specific category of those that make up the variables age, level of education and employment status. This operation was repeated for each of the categories of these three variables, implementing a total of 84 ANOVA tests.

For each of these analyses, a Levene variance homogeneity test was performed. According to the results of this test, the general goodness of fit of the model was observed from F-test of the ANOVA, in the case of homogeneous variances, and based on the Welch's statistical analysis, in the case of non-homogeneous variances. All the analyses implemented were significant.

Subsequently, in order to be able to compare percentages, including the different categories of the three socio-demographic variables and the year of survey, the Tukey's indices were used in the case of homogeneous variances, and the Games-Howell indices were used in the non-homogeneous variances. Given that the size of the samples taken in the different years is not the same, the analysis was implemented using harmonic averages of the sizes of the samples. For reasons of space, not all the indices of the different analyses implemented can be presented here. As a result, we shall only mention those differences in percentages which, based on the ANOVA tests, turned out to be significant and those that did not.

10.4 Results and Analysis of Results

10.4.1 Access

Figure 10.2 shows the percentages of Internet access in Spain between 2004 and 2011. All these percentages are significantly different from one another. In other words, there is a statistically significant increase in the use of the Internet in Spain during the period 2004–2011. Specifically, growth amounted to almost 27 percentage points, going from an Internet access percentage of 40.4 % of the Spanish population (2004) to 67.1 % (2011).

As mentioned, in order to assess the existence and the evolution of the Digital Divide in Spain, we should consider whether this evolution has the same patterns across the socio-demographic variables of age, level of education and employment status.

Figure 10.3 shows the evolution of Internet access by different age groups. The results of the ANOVA tests show that for the different categories of the age variable, and in each of the years, Internet access percentages are significantly different. If we look at the last year of the series (2011), the data show a consistent difference between the levels of access reached in the two youngest cohorts of population, respectively, representing percentages of 95 and 87.8 %. Also significant are the differences between the two oldest segments of population. However, the level of penetration of the Internet in these groups is considerably lower, at 15.6 and 37.7 %, respectively. In addition, these two latter age groups show a slower growth rate. Specifically, 24 % (55–64 years old) and 12.6 %

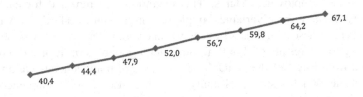

| | 2004 | 2005 | 2006 | 2007 | 2008 | 2009 | 2010 | 2011 |

Fig. 10.2 Internet access penetration percentages (2004–2011). *Source* INE

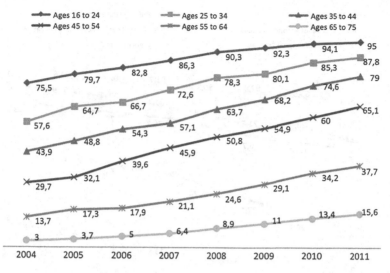

Fig. 10.3 Internet access penetration percentages by age (2004–2011). *Source* INE

(65–74 years old) compared to, for instance, 32 % (25–34 years old) and 36 % of growth in the groups of 35–44 years old and 45–54 years old.

Thus, we can see two social groups forming within the age variable, depending on whether people are older or younger than 55. Whereas among those older than 55, the percentage of Internet use penetration never reaches 40 %, and among those younger than 55, this percentage is close to, or even surpasses, 70 %. Likewise, as has been said above, the percentage of evolution in Internet access of both cohorts in the period described is higher among the latter than among the former. Thus, given the level of Internet use penetration among people older than 55 and the rate of recruitment of new users, it is difficult for the Digital Divide to reduce, given that this group cannot be expected to reach, according to the data we have at present, a level of saturation as high as that of the younger age groups.

Naturally, we might speculate with the idea that in some years' time, when people who are now older than 55 have died, this type of Digital Divide shall not have the effect it has at present. This is one of the ideas the thesis of *normalisation* has been based on. However, even if this were true, as we shall see in what follows, this thesis could not easily be applied to other variables such as level of education or employment status (Fig. 10.3).

Figure 10.4 shows the evolution of population percentages in Internet access according to the level of education of Spanish citizens.

Also in this case, the results of the ANOVA tests show Internet access percentages that are significantly different across the different categories and each of the years. The chart shows how population segments with the highest level of education (higher education, second level of secondary education and higher professional training) show higher Internet access penetration rates (95.1, 90.5 and 83.3 %, respectively). The growth rate of subjects with higher education is the

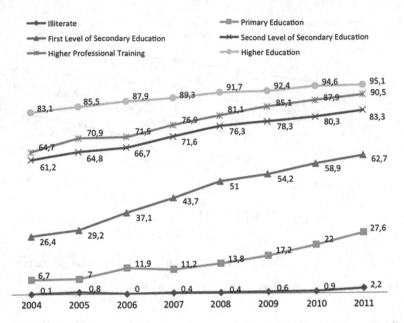

Fig. 10.4 Internet access penetration percentages by level of education (2004–2011). *Source* INE

lowest of these three segments, with 12 %, although the percentage that started the series (2004) was already considerably high (83.1 %). On the other hand, people with an education level of second level of secondary education and higher professional training show high growth rates (25.5 and 22.1 %, respectively), having a much lower starting point.

People who have completed the first level of secondary education also show a high growth rate, of around 36 %. In fact, it is the group with the highest growth rate in recent years. Even so, their access percentage in 2011 is still around 20 to 33 percentage points below that of the three population groups with the highest levels of education.

Those with primary education show a significant growth rate (20.9 %). However, given that their starting point in 2004 was a very low penetration percentage (6.7 %), by 2011, they had only reached a rate of 27.6 %. Lastly, we should note the almost flat growth rate of access percentages of people with no formal education.

Our analysis allows us to observe two groups within the variable "level of education": on the one hand, people with a very low or non-existent level of education and, on the other hand, people with middle and higher education. Among the latter, as we have said, the level of Internet use penetration is very high. This is due either to a high growth rate or to the high level of Internet use penetration in the year the available series started. Currently, both due to the current Internet penetration percentage and due to the rate of incorporation of new

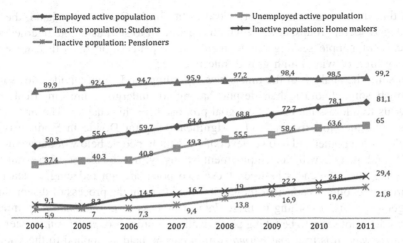

Fig. 10.5 Internet access penetration percentages by employment status (2004–2011). *Source* INE

users among less educated people, it is not possible to hold that there is a convergence process with the groups with higher levels of education, nor that there is a possibility of them reaching high levels of saturation. In this regard, to analyse the Digital Divide requires thinking in terms of the polarisation that is generated between those more and those less privileged in terms of education.

Lastly, fig. 10.5 shows the Internet access penetration rates across the last eight years, among population segments characterised by the different employment statuses.

The ANOVA tests implemented for this case show how, among the different categories of the employment status variable and in each of the years, all the Internet access percentages are significantly different from the others. The percentages shown in this figure show how students are the group with the highest level of Internet penetration, reaching a rate of 99.2 % in 2011. Employed and unemployed active population also shows high Internet access percentages. Employed people show similar rates to that of the students and show an ever higher growth rate (30.8 %). The unemployed also show a high growth rate (27.6 %), but their relatively low starting point in 2004 (37.45) means that they are still at a considerable distance (34 % points) from students.

The groups made up of pensioners and homemakers represent a separate set of people. The two groups show the lowest Internet access percentages, relatively close across all the years of the series. As to the growth rate, they are also similar, with 20.3 and 16 %, respectively. In other words, these are growth rates, compared to the employed and unemployed active groups, respectively, of almost a third and half.

Our hypothesis that the study of the Digital Divide should be analysed as a process of polarisation between the more and the less privileged social groups is clearer in this variable. The low growth of pensioners and homemakers leads us to hold that their saturation rates are close to the limit. On the other hand, the less privileged social groups, or those with more stimuli to start using the Internet, as is the case of people seeking employment, show penetration rates that are either already high or with a high growth rate Fig. 10.5.

In sum, the analysis of the time series regarding the Digital Divide allows us to establish some findings that, despite having to undergo future empirical tests, allow us to address the theoretical goal proposed for this chapter. The first is that there continues to exist in 2011 a significant Digital Divide in Spain between people with a higher and a lower level of education, people below or above the age of 55 and people with an employment status with a better or worse outlook. Likewise, the differences between these two poles are not reducing, at least not vigorously enough to consider the Digital Divide as in the process of disappearing altogether. Thus, according to these data, a process of *normalisation* in Internet access does not seem to be taking place. Rather, a process of *stratification* seems to be under way. It is true that *normalisation* can be held as applied to the younger groups, the most educated groups and people with the most favourable employment statuses. That is, the differences between the groups with greater levels of education with a better employment status and among the young have indeed reduced. However, the great social differences, far from disappearing, are evident in our study. This is the reason why we put forward the thesis that there is a process of normalisation among the more privileged groups at the same time as there is a process of stratification between the more and the less privileged. Our interpretation of this result is addressed in the conclusion.

10.4.2 E-Shopping

Chart 10.6 shows us the trend in purchasing products or services online in Spain in the years included in the series. It should be noted that these percentages do not refer to the Spanish population, but to Spanish Internet users. The ANOVA tests have shown significant differences between the percentages of all the years that make up the time series, including between years 2007 and 2008. Despite the trend slowing down in these two years, the trend shows a stable and significant growth process across the entire period (fig. 10.6).

However, it is important to know whether said evolution has occurred homogeneously or not, that is, across all population groups. Figure 10.7 shows e-shopping penetration percentages among Spanish Internet users according to the age variable categories.

According to the ANOVA tests performed, all the percentages of the categories have turned out to be significantly different from each other in all the years of the series. The data show Internet users from 25 to 34 years of age as those more likely to be e-shoppers (30.2 % in 2011). Their growth rate is also the highest, with a 20.3 % increase since 2004. Internet users in the groups of 16–24 years old and 35–44 years old show similar percentage and growth rates. Both segments started

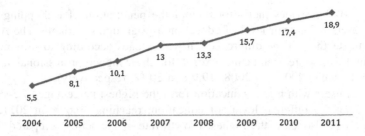

Fig. 10.6 E-shopping penetration percentages (2004–2011). *Source* INE

with e-shopper percentages of around 6 % of Internet users in 2004 and reached
21–23 % in 2011.

In 2004 and 2005, at the start of the series, the groups from 45 to 54 years of
age and 55–64 years of age showed similar e-shopping penetration percentages.
However, from 2006, the two age groups show differently evolving profiles. Thus,
whereas Internet users from 45 to 54 years of age reached percentages of 16.2 %
in 2001, those from 55 to 64 only reached 9.2 %. As was to be expected, the group
with less Internet users prone to engaging in e-shopping is that comprising people
from 65 to 74 years of age.

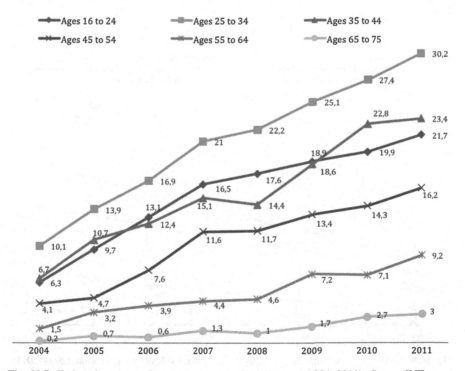

Fig. 10.7 E-shopping penetration percentages in Spain by age (2004–2011). *Source* INE

Figure 10.8 provides indications about the penetration of e-shopping among Internet users using their level of education as a grouping variable. The ANOVA tests indicate that all the differences in percentages according to different categories and years are significant, except for the "higher professional training" category between 2007 and 2008 (19.9 and 20 %, respectively).

Internet users with higher education form the highest percentage of e-shoppers from among the different levels of education, reaching 39.6 % in 2011. These Internet users also show the highest growth rate of the series compared to other levels of study (23.1 %). On the other hand, Internet users who have completed the second level of secondary education or higher professional training show similar e-shopping percentages up to 2006. Subsequently, both groups grow, though Internet users who have completed the second level of secondary education grow at a faster rate (30.5 % in 2011) compared to those with higher professional training (21.7 % that same year). Internet users who have completed the first level of secondary education show a more marked pattern of growth as from 2006, and despite not reaching the percentages that characterise higher levels of education, they represent 10.8 % of e-shoppers in 2011.

Both Internet users with primary education or with no formal education show extremely low e-shopping growth rates. Nevertheless, those with primary education represent 2.9 % of penetration in 2011, although those lacking in formal education altogether continue showing percentages of around 0 %.

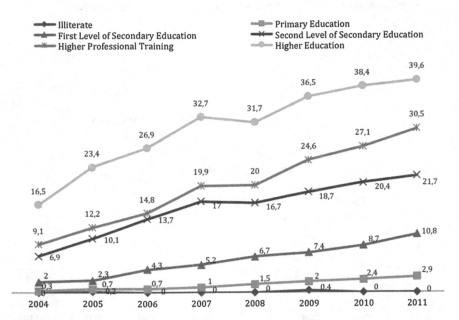

Fig. 10.8 E-shopping penetration percentages in Spain by level of education (2004–2011). *Source* INE

Figure 10.9 contains the percentages of e-shopping penetration based on the employment status of Spanish workers. Again, the ANOVA tests carried out show in this variable that all the percentage differences according to the different categories and years are significant, except for the categories of pensioners and homemakers in 2007.

This figure shows the evolution of e-shopper percentages following three different patterns. The first pattern includes student and active population Internet users and is characterised by having the highest e-shopping rates (25.8 and 25.4 %, respectively). In addition, both groups show the highest growth rate, with a 17.9 % growth across the period of study.

The second pattern is that defined by unemployed Internet users, which reached in 2011 an e-shopping percentage of 12 %, showing a growth rate of almost 10 %. And, lastly, the third growth pattern concerns pensioners and homemakers, who show the lowest e-shopping percentages, as well as lower growth percentages. Pensioner growth is, during the period studied, 5.2 percentage points and that of homemakers, 3.2 points.

To sum up, it can be concluded that at the start of our time series, a similar or not-too-distant starting point could be seen among the different social groups studied according to the different socio-demographic variables examined. In our opinion, this is due to the fact that the starting point of the series is very close to the start of this kind of digital behaviour. However, the most privileged groups immediately take the lead on the uptake of this activity, thanks to their greater inter-annual growth. This happens more markedly among people aged 25–34, among students and people who are active workers and among people with university studies. It can also be seen that young people with higher levels of

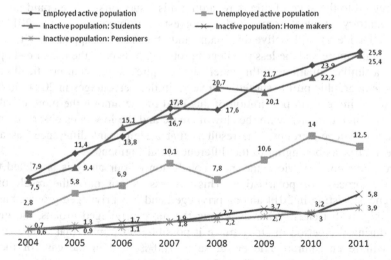

Fig. 10.9 Percentages of e-shopping practice in Spain by employment status (2004–2011). *Source* INE

education show a high growth rate, allowing them to approach the percentages reached by the most privileged groups. On the other hand, the groups formed by older people, with lower levels of education or in a less privileged situation in the job market, are left behind in the process of incorporation to this kind of digital activity.

The data allow us to hold that Digital Inequality, understood as a consequence of e-shopping, is increasing. The potential benefits of this kind of Internet use are being harnessed to a greater degree by privileged groups as predicted by the literature. However, it is important to point out that there is a significant polarisation between the more and the less privileged. Thus, Digital Inequality has an effect on large groups within every socio-demographic variable.

Taking into account the low growth rate observed among people with primary education or without any formal education, and among retired people and homemakers, we could put forward the theory that high saturation levels will be reached in the midterm. This makes us think of the difficulty of a process of normalisation in the use of e-shopping. Therefore, our theory is that, according to the current development of the penetration of this kind of digital practice, we find ourselves facing a process of *stratification* of the differences between those that are more socially privileged and those that are less socially privileged.

10.5 Conclusions

Our study on the Digital Divide in Spain, as well as on a representative type of BAUI, e-shopping, has allowed us to offer certain conclusions regarding the phenomenon of Digital Inequality that are relevant to furthering our knowledge with regard to the state of affairs regarding this issue. Firstly, we could point out that, contrary to what some studies suggest (Eurobarometer 2002–2011), the Digital Divide is still active in Spain, and, far from disappearing, it persists between the more and the less privileged groups. On its part, the use of e-shopping shows a similar situation. In this latter case, despite the fact that practically every socio-demographic profile started with very similar percentages in 2004, there has been a fast increase in penetration of this kind of use among the more privileged citizens which contrasts with the slow increase and the low rates observed among the less privileged groups. The result is that a significant difference has arisen between citizens belonging to the different social segments.

From our point of view, the main conclusion of our analysis is related to an apparent process of polarisation. This process is not only the result of the differences existing in 2011 among privileged and less privileged groups, but also of their growth trends. Whereas among the more privileged groups, the growth rates during the period of study make it possible to estimate that in near future, they will reach saturation levels close to 100 %, and in the case of the less privileged groups, it will be difficult for said saturation rate to reach the total of the respective group.

This leads us to estimate that both in the case of the Digital Divide and in the case of e-shopping, no normalisation process is taking place. The less privileged groups will find it hard to reach, with the growth percentages observed, the percentages observed in privileged groups. Therefore, we consider that, in general terms, we find ourselves immerse in a *stratification* process. This is because the privileged and less privileged groups represent different saturation levels.

Thus, the penetration of the use of the Internet, as well as the studied case of BAUI, reproduces the existing inequalities in general terms among people with more resources and people with fewer resources. According to this reading, citizens who are better-positioned are those who will have more opportunities to benefit from the advantages offered by this medium. In the case of e-shopping, we would be dealing with a case of Digital Inequality inasmuch as the citizens with more resources would be in a better position to access services and products offered online, whereas citizens with fewer resources would have fewer possibilities of accessing those advantages. We would therefore be talking about a Commercial Digital Divide.

In short, for the Spanish case study, our analysis puts forth a scenario of Digital Inequality both in the use of the Internet and in the use of one of the advantageous services provided by this medium. This scenario cannot only be perceived in the 2011 data, but based on the time series studied, it cannot be said that it will decrease in the future. This is why we must remember that the Information and Knowledge Society poses a risk of polarisation among the more and the less privileged citizens which, based on the data exposed here, is very present in the development of said society in Spain.

References

Attewell P (2001) The first and the second digital divides. Sociol Educ 74:252–269

Bimber B (2000) Measuring the gender gap on the internet. Soc Sci Q 81:868–876

Bonfadelli H (2002) The internet and knowledge gaps. A theoretical and empirical investigation. Eur J Commun 17:65–84

Cooper J, Weaver K (2003) Gender and computers: understanding digital divide. Lawrence Erlbaum Associates, Mahwah

DiMaggio P, Hargittai E (2001) From the digital divide to digital inequality. Studying internet use as penetration increase. Centre arts Cultural Policy Stud 15:1–23

DiMaggio P, Hargittai E, Neuman WR, Robinson JP (2001) Social implications of the Internet. Ann Rev Soc 27:307–336

DiMaggio P, Hargittai E, Celeste C, Shafer S (2004) From unequal access to differentiated use: a literature review and agenda for research on digital inequality. In: Neckerman KM (ed) Social inequality. Russell Sage Foundation, New York, pp 355–400

Eurobarometer (2002–2011) http://epp.eurostat.ec.europa.eu/statistics_explained/index.php/Information_society_statistics

Grewal D, Levy M (2004) Internet retailing: enablers, limiters and market consequences. J Bus Res 57:703–713

Gunkel DJ (2003) Second thoughts: toward a critique of the digital divide. New Media Soc 5:499–522

Hague BN, Brian DL (1999) Digital democracy: discourse and decision making in the information age. Routledge, New York

Hargittai E (2010) Digital na(t)ives? Variation in internet skills and uses among members of the "net generation". Sociol Inq 80:92–113

Hoffman D, Novak, TP, Scholosser AE (2001) The evolution of digital divide: examining de relationship of race to Internet access and usage over time. In: Compaine BM (ed) The digital divide. Facing a crisis or creating a myth? The MIT Press, Cambridge, pp 47–99

Horrman DL, Novak TP (1998) Bridging the racial divide on the internet. Science 17:145–164

Levy P (2002) Cibercultura. La cultura de la sociedad digital, Antropos, Madrid

Negroponte N (1996) Being digital. Vintage, London

Norris P (2001) Digital divide? Civic engagement, information poverty and the Internet worldwide. Cambridge University Press, Cambridge

Nozick R (1991) Philosophical explanations. Harvard University Press, Harvard

NTIA (National Telecommunications and Information Administration) (1999) Falling through the net II: defining digital divide

NTIA (National Telecommunications and Information Administration) (2000) Falling through the net II: toward digital inclusion

NTIA (National Telecommunications and Information Administration) (2002) A nation online: how Americans are expanding their use of the internet

Robles JM, Torres C, Molina O (2010) Brecha digital. Un Análisis de las Desigualdades Tecnológicas en España. Sistema 218:3–22

Rockwell S, Singlenton L (2002) The effects of computer anxiety and communication apprehension on the adoption and utilization of the Internet. J Commun 12:123–145

Rojas V, Straubhaar J, Roychowdhury D, Okur O (2004) Communities, cultural capital and the digital divide. In: Bucy E, Newhagen J (eds) Media access: social and psychological dimensions of new technology use. Lea, London, pp 107–130

Strover S (1999) Rural Internet connectivity. Rural Policy Research Institute, Columbia

Torres-Albero C, Robles JM, Molina O (2011) ¿Por qué usamos las Tecnologías de la Información y las Comunicaciones? Un estudio sobre las bases sociales de la Utilidad Individual de Internet. Revista Internacional de Sociología 69:371–392

Van Deursen AJ, Van Dijk J (2009) Improving digital skills for the use of online public information and services. Gov Inf Q 26:333–340

Van Deursen AJ, Van Dijk J (2010) Internet skills and the digital divide. New Media Soc 12:1–19

Van Dijk J (2005) The deepening divide. Inequality in the Information Society. Sage Publications, California

Van Dijk J (2006) Digital divide research, achievements and shortcomings. Poetics 34:221–235

Walsh E (2000) The truth about the digital divide. Forrester, Cambridge

Wang F, Head M, Archer N (2002) E-tailing: an analysis of web impacts on the retail market. J Bus Strat 19:73–93

Chapter 11
From "Singularity" to Inequality: Perspectives on the Emerging Robotics Divide

Antonio López Peláez and Sagrario Segado Sánchez-Cabezudo

11.1 Introduction

As we saw in the introduction, industrial robots are a mature and competitive product. According to the International Federation of Robotics (IFR), in 2011, a record number of industrial and service robots were installed: "In 2011, robot supplies increased by 38 % to 166,028 units, by far the highest level ever recorded for one year. The value of sales surged by 46 % to US$8.5 billion, a new record. It should be noted that this value generally does not include the cost of software, peripherals and systems engineering. Including the mentioned costs might result in the actual robotic systems market value to be about three times as high. The worldwide market value for robot systems in 2011 is therefore estimated to be $25.5 billion. In 2011, Japan was again the biggest robot market in the world. Robot supplies to Japan continued to recover and increased by 27 % to almost 28,000 units. The IFR Statistical Department expects that between 2013 and 2015 worldwide robot sales will increase by about 5 % on average per year. In 2015, the annual supply of industrial robots will reach more than 200,000 units" (IFR 2012a).

Industrial restructuring and the reorganization of the service sector have gone hand in hand with an increase in the installation of automated and robotic systems. The effects of automation have gone beyond the optimization of goods or services production systems to change the world of work and leisure. Automation has also had a clear impact on economic development and the power associated with it as technological innovation and competitiveness are linked to the intensive use of robots and automated systems for a wide range of purposes. In this regard, and in

A. López Peláez (✉) · S. Segado Sánchez-Cabezudo
Department of Social Work, Faculty of Law, National Distance Education University (UNED), C/Obispo Trejo 2 28040 Madrid, Spain
e-mail: alopez@der.uned.es

S. Segado Sánchez-Cabezudo
e-mail: ssegado@der.uned.es

A. López Peláez (ed.), *The Robotics Divide*,
DOI: 10.1007/978-1-4471-5358-0_11, © Springer-Verlag London 2014

relation to the possible dimensions of what we call the robotics divide, we must take into account the evolution of military and space robotics. Defense robotics, for example, is already the largest market for service robots today: "About 16,400 service robots for professional use were sold in 2011, 9 % more than in 2010. With about 6,600, service robots in defense applications accounted for 40 % of the total number of service robots for professional use sold in 2011. Thereof, unmanned aerial vehicles seem to be the most important application as their sales increased by 11 % to more than 5,000 units. The value of defense robots can only roughly be estimated. It was about US$748 million, 3 % higher than in 2010" (IFR 2012b).

Throughout the chapters of this book, we have analyzed the evolution and characteristics of advanced robotics and the areas in which this technology is expected to expand while exploring a key issue: the socioeconomic model in which this technology is being developed and incorporated. To do so, we have taken as a reference the most recent technological divide occurring in advanced societies: the digital divide. In this last chapter, we present the final results of prospective research we have conducted over the last ten years on robotics and advanced automation,[1] focusing on the issues that we consider key in any technological revolution: the transformation of the world of work and leisure, military power, and the characteristics of a society that in the next 20 years will no doubt be marked by the intensive development of robotics technology.

11.2 Why are There Technology Divides? Opening the Black Box of Scientific and Technological Innovation

The question that gave rise to this interdisciplinary book—are we witnessing the beginning of a new digital divide, the divide robotic?—can only be asked following years of research that has highlighted the key role technology has played in the dynamics of power and social inequality in societies throughout history. But it also arises from research showing the extent to which societies develop and apply technologies to reinforce the status quo and consolidate the distribution of power

---

[1] To do so, we conducted four consecutive Delphi studies with experts in information and communication technologies, robotics and genetic engineering, and biotechnology in 1996, 2002, 2005, and 2010. Fifty experts were selected from each of the three research lines to participate in the four waves; hence, we interviewed a total of 600 experts using the Delphi method. In each of the four waves, we conducted a preliminary study and administered a pilot questionnaire and a final questionnaire in two rounds. Some of the results obtained in the area of robotics, which was coordinated by Professor Antonio López Peláez, have been referenced in the OECD Future Studies database and published in leading journals in the field of technological forecasting, among them *Technological Forecasting and Social Change* (Lopez Peláez and Kyriakou 2008), *Social Epistemology* (López Peláez and Díaz Martínez 2007), *The IPTS Report* (López Peláez and Krux 2000, 2002, 2003), and *Robotics* (López 2000).

between individuals, groups, communities, and states. The difference with other periods of history lies in a key issue which could also be considered "organizational technology," namely democracy. In societies where people wish to be masters of their own destiny, the question arises as to the vested interests in technological developments as well as the more or less bloody battle that is fought between various actors to achieve hegemony in a given field of technology, be it with regard to enterprises, patents, or geostrategic conflicts. Awareness of the social and political nature of technological development has occurred in two phases. In a first phase, the technological model and investment priorities took center stage in a historical context in which science policy became a strategic issue. In a second phase, the redefinition of old forms of social exclusion and the emergence of new forms of social exclusion associated with new technologies emphasized the need to address the intended or unintended consequences of technological development.

Three main approaches are taken to explore these issues: (1) prospective studies and trend analyses which examine technologies considered crucial for the immediate future and which provide information for decision making; (2) case studies, which highlight the characteristics of a particular sociotechnical model of development; and (3) research to assess technology, scientific and technological literacy and science policy, and examine the social and participatory mechanisms that permit citizens as political subjects, consumers, or stakeholders to participate in designing their own social model. In our opinion, it is important to analyze the forecasts of experts, as they provide insight into the probable scenarios and the social model in which technological advances will be integrated in the future. For this reason, in our prospective research, we have always examined key issues concerning the labor market, work practices, the expansion of service robotics (including military robotics), and the main effects or consequences in the coming years of technological innovations and the social model in which they are embedded; a model which, in turn, is already structuring such innovations.

In the sociology of science and technology, there are two main lines of research on the present and future evolution of technological innovations, and their dialectical relationship with the society that creates them; a society which is both reflected in and transformed by these innovations in often unexpected ways. One line of academic research is dedicated to what are known as sociotechnical systems in which users play a major role (Bijker 1995). Constructivist approaches have highlighted the theoretical inability of deterministic explanations to give an account of the process of innovation and technology development we create and which shapes our lives. In designing a technological system, users can become genuine counter-programmers and unexpectedly transform the expected result. For example, the introduction of information and communication technologies in the public administration can be used to reinforce the hierarchical structure of an organization rather than achieving the theoretical objectives underlying their implementation such as greater flexibility and increased productivity (Aibar 2011).

The second line, which is linked to a strong tradition of applied research in the field of technology planning, has led to the creation of research institutes within and outside our borders which analyze and forecast the evolution of technologies with a strategic purpose, and whose approach is also sociological. The aim of these institutes is to obtain qualified information in order to provide support to the companies and institutions that fund the research in deciding where to invest their scarce economic resources. Institutes such as the Industrial Prospective Technology Observatory (OPTI) in Spain and the Institute for Prospective Technological Studies (IPTS) in the European Union share a common theoretical orientation: science policy as a strategic activity. In recent decades, these institutes have developed research programs on technological evolution from a sociological perspective in which they increasingly analyze user behavior with the aid of experts in technological forecasting.

In our opinion, constructivist approaches have led to significant changes in the methods and theoretical models of what might be called prospective sociology and in sociological research dedicated to the study of social trends and their evolution in coming years. Prospective studies no longer use only the opinions of technology experts defined in a very narrow manner in interviews, such as the Delphi method, or in discussion groups. Given that users play a key role, and the production of technology is a social process subject to the constraints and opportunities of a given socioeconomic environment, it is important to take into consideration the informed judgments of experts other than engineers or scientists who investigate a particular technology. We must also include other kinds of experts: those who influence, consume, and appropriate technological innovations, those that can give an expert opinion on the development of a technological innovation, its final characteristics, and how it can be shaped (and in turn shapes us) according to the social logic in which it is embedded and in which it is produced (and transforms). This has led to the redefinition of the figure of the expert, which also includes users, and in many cases early adopters-cum-epistemological leaders (López Peláez and Díaz Martínez 2007).

This reorientation of prospective studies could be described as a process that evolves from the analysis of the impacts of technology (based on forecasts by technology experts in the strict sense) to the analysis of the sociotechnical design of technology (based on forecasts by experts who are not only engineers or scientists but also economists, designers, skilled users, etc.). On the one hand, this redefinition of technological trajectories (and their relationship to the society that creates them and is transformed by them) has made the process to select experts more difficult; a process which is key in prospective methodologies. On the other hand, it allows us to respond better to basic questions in prospective studies, which also have a political, economic, or marketing aim such as the competitiveness of a country, or legal regulations (on designs, electrical systems, etc.) regarding a particular innovation to ensure its successful integration in domestic environments, all of which are key aspects dealt with in Delphi studies conducted in Japan.

Technology scholars have drawn quite similar conclusions from very different theoretical approaches. The analysis of Collingridge (1980) on technological development showed that once a technological model is established, it tends to remain stable by its own inertia and assimilate innovations within its operating logic, while Orlikowski (2000), in discussing what she calls technologies-in-practice (TIP), highlights that the interaction between ICTs and organizations is likely to follow an inertial development model. Technology serves to maintain existing users (and we must remember that we are immersed in a sociotechnical system that is sometimes invisible because we do not reveal it, but which previously establishes how to incorporate technologies). In our opinion, our neoliberal and competitive model is a form of organization that integrates advanced robotics. Beyond romantic expectations about a happy future shared by humans and machines, we must consider the inertia (guided by the logic of the market) that will shape the incorporation of industrial and service robotics. An inertia, which in line with Orlikowski, may reinforce the status quo, and strengthen existing hierarchical structures. In our neoliberal economic environment, the result is clear: It is highly probable that the processes of social and economic segmentation will be fortified by the incorporation of robotics in increasingly more spheres of social life. This hypothesis, the emergence of a new social divide that transcends the digital divide, what we call the robotics divide, is largely shared by the experts who participated in our research.

Although robots have long been a dream for humans, in the last decades of the twentieth century, the fantasy has been fulfilled as robots have become commonplace in industry, and begun to expand rapidly in the service sector. Interactions between humans and machines are one of the most fascinating areas of research in the social sciences. Indeed, robots, those fairly intelligent machines, are present in our collective imagination, whether it be in sci-fi movies, day-to-day life, or at the workplace. From a perspective that seeks to achieve a balance between technological forecasting related to innovation, development and the competitive edge of countries, and constructivist approaches emphasizing the social dynamics that influence and shape the process of designing and integrating technologies, we have redefined the concept of "expert." Specifically, we have selected a panel of experts which includes engineers, designers, experts in the implementation of robots, or early users in the field of service robotics.

If we define technology as a fact and a social process, trend analysis permits us to achieve three basic objectives to intervene in shaping the technological model that is part and parcel of our society. First, to anticipate likely outcomes; second, to open, in Bourdieu's terms, "the space of possibles"; and, thirdly, to analyze the sociotechnical model in which technology is designed and integrated in our cosmopolitan society. To do so, it is necessary to consider the forecasts not only of the experts who design technologies, but also of the engineers who apply them, the managers of companies that install them, and those who use them. In this regard, our approach is innovative insofar as it broadens our understanding of the processes of change in which we are immersed. Our approach overcomes the limitations of traditional prospective studies which analyze the impacts of innovations (and often presuppose an approach based on the theoretical assumptions of technological determinism) as it includes basic issues in any constructivist approach, such as users' interests or social models in which technology is developed (and where technological innovations can reinforce systems of social stratification and preexisting inequality). The methodology we use in the prospective studies we are conducting on robotics and advanced automation, as well as some of the results presented in this chapter, is based on the analysis of technological trajectories, and research on the sociotechnical model in which they are developed and applied, introducing issues other than the mere forecasting of events. International comparative studies on research in this area have shown the need to pursue this line of analysis, corroborating the methodology we have used to date (COTEC 2003; NISTEP 2005: 2).

A large number of methodologies are used in what could be called, in the broad sense of the term, futures studies: the analysis of key technologies, technology roadmapping, scenario development, and a long etcetera (Keenan et al. 2006: 3–6). In our research on the evolution and impact of robotics and advanced automation, we used the Delphi method for three reasons:

- First, regardless of the debate on the weaknesses and strengths of the Delphi technique (Goldschmidt 1975), it is now a consolidated method (Linstone and Turoff 1975) that is widely used in science and technology in general (Hader and Hader 2000). The Delphi method has also been widely used in research on social trends from the 1970s onwards (Adler and Ziglio 1996).
- Second, important studies on technological forecasting conducted by European, Japanese, and American research institutes have used this method (MESR 1995; Loveridge et al. 1995; ITK 1998; FISI 1998; NISTEP 2002, 2005; COTEC 2006).
- Third, group interaction on which the Delphi method is based is particularly suitable for revealing forecasts on technology implementation, the theoretical models that guide the interaction of technology, the organizations and the societies in which it is developed, and also the reflective views of experts which often do not coincide with the linear and optimistic forecasts of the interest groups that fund and promote the uncritical incorporation of technologies (assuming, in the best tradition of technological determinism, that the innovation itself will solve any problem of implementation). For our study, we selected a group of experts from the field of technological innovation; managers in highly automated companies; robot and automated systems software programmers; engineers and technicians who work, install, and live with automated and robotic systems; as well as union representatives of companies with high levels of automation and early users of robotics services (managers of hotels that use service robots, users of cleaning robots, etc.).

The design, setup, and practical incorporation of technology are a complex social process, and the forecasts and discourse of experts provide us with clues to develop trend scenarios from which we can open the space of possible, especially when technological innovations may affect our very survival as a species. Although the forecasts of experts are not the only factor involved in developing technologies, they are an important element in the process of technological innovation, and the social construction of our future as they permit us to visualize the development model in which we are immersed and the foreseeable sequence of innovations depending on the state of the discipline in question, the demands of the production model and the provision of services for which such technologies are to be used. In our opinion, the forecasts of experts permit opening the black box of technology, especially when the panel is selected in a proper manner.

Prospective studies should combine several methods and give a voice to the groups involved in the development and implementation of technologies (as Professor Eduard Aibar shows in this book). In a recent study, we combined the Delphi method with the study of histories and in-depth interviews to analyze the consequences of rail transport liberalization in Europe (given that "organization" is a form of "technology"), and its foreseeable consequences on occupational safety and health in the sector (López Peláez and Segado Sánchez-Cabezudo 2009; López Peláez and Segado Sánchez-Cabezudo 2010; López Peláez et al. 2012). There are other research methods that can also serve as a complement to prospective research, such as case studies, group discussions between users affected by a particular technology, or the analysis of political culture and institutions. Given that the research on service robotics is still in the early stages of development, we compared the results of our study with those of the most important research recently published in the field (Kurzweil 2005), which employs other methodologies. In doing so, we hope to contribute to the debate on the present and future of technology from a sociological perspective that transcends technological determinism, but that is not strictly constructivist. Our aim is to find a balance between theory and method in which the figure of the expert is redefined while seeking to enrich our understanding of a probable future and above all to highlight the emergence of a new technological divide: the robotics divide.

11.3 Seeking the Origin of the Robotics Divide: Forecasts on Technological Development by the Year 2020

As noted in the second chapter of this book, it is important to distinguish some key points in the analysis of what we call the *robotics divide*, and which share features in common with other segmentation and stratification processes resulting from previous technological revolutions:

• First, the very nature of technology, its "identity" or "internal logic," and its performance requirements (energy, organizational, etc.) are a key issue in robotics as the robots of the future are designed to be intelligent machines capable of learning. Indeed, leading researchers such as Kurzweil foresee the emergence of a new subject, an alter ego, which may one day claim its own rights. Beyond the essentialist debate on this issue, we must explore the development of robotics technology for purposes of work, entertainment, and companionship as Levy notes, or as a combatant in the battlefield. As we will see below, the main predictions of the experts interviewed are the trend toward a greater *hybridization between humans and robots* due to the similarity and

proximity of the capabilities of robots and humans, as well as the progressive implantation of robotics technology in our bodies (and the emergence of a new robotic-human subject, the cyborg).
- Second, it is of interest to analyze how robotics is transforming the world of work and leisure, the workplace and the home, and explore in-depth the socioeconomic model that, according to the experts, shapes our technologies. From this point of view, future trends underscore present commitments. As has occurred with other technologies (López Peláez et al. 2012), our socioeconomic model does not appear to transform itself to exploit the potential of robotics, but is incorporated as a technology, reinforcing existing power structures.

The results of our prospective research in the last 10 years can be examined through the lens of these two analytical spheres. First, the evolution of technologies needed to develop "more human" robots (touch, vision, mobility, speech, intelligence as defined in Table 11.1), and the extent to which such advances coincide with the forecasts of other international research institutes. Second, the model incorporating this technology: what level of automation is expected to be achieved, how will it affect current working practices, and to what extent is it going to become a staple in the world of work and leisure in the next decades. According to the experts, the discussion of technological advances is a relevant issue for society as a whole due to foreseeable widespread incorporation of service and domestic robotics in the medium term, some 10 years. As we have shown in the previous sections, our ability to determine and set the path of technological development is greater when we are in the early stages of development, and decreases once the technology is finally implemented.

11.3.1 An Ever Closer "Singularity"?: Innovations in Robotics Technology in the Coming Years

According to the forecasts of experts, the advances that will permit the intensive and coordinated use of vision, intelligence, language, and navigation technologies will occur in the second decade of the twentieth century. The forecasts fall into two different but complementary spheres: forecasts on the technological advances needed to achieve these technologies, and the incorporation of robots with these capabilities in specific areas. In line with these forecasts, we conducted a series of Delphi surveys up to the year 2005, which we complemented with further questions that we asked the experts in 2011.

- Industrial robots with 3D vision are expected to become commercially available (Table 11.1) by the year 2015. The current limitations in the fields of intelligence, mobility and integration in a natural language system are expected to be overcome by the year 2020, while robots are expected to integrate these technologies and significantly increase the ability to automate tasks in all areas of activity 5 years later, in 2025. The forecast that the different technologies will

Table 11.1 Forecasts on advances in robotics technology by the year 2020

Events	Delphi study 1996		Delphi study 2002		Delphi study 2005	
	Occurrence horizon	Forecast certainty	Occurrence horizon	Forecast certainty	Occurrence horizon	Forecast certainty
"Intelligent" industrial robots will be available (intelligence refers to the ability of robots to make decisions regarding the tasks performed when the robots have been previously trained to perform such tasks, and use information from the environment to make decisions)	2010	4	2014	4	2020	3
3D vision robots will be available	2010	4	2015	3	2015	3
Robots integrating natural language systems will be available	2010	3	2025	3	2020	3
Robots capable of moving in previously unknown environments will be available	2010	4	2020	3	2020	3
Robots integrating the four above capabilities will be available	2025	3	2025	3	2025	3

Note Forecast certainty: 5 degrees (*1* uncertain, *2* not very certain, *3* certain, *4* quite certain, *5* very certain). *Source* Working Group on Social Trends (GETS 1996, 2002, 2005) be fully integrated by 2025 is in line with the forecast made by the experts in the Delphi study we conducted in 1996 and 2002, indicating that it is a relatively stable temporal prediction that has not been delayed. However, we should note that in the case of technological developments related to the "intelligence" of robots, there is a ten-year lag between the forecast on the availability of robots with this ability, specifically from 2010 (Delphi study of 1996) to 2020 (Delphi study of 2005).

- In the most recent Delphi survey we conducted in 2011, we presented the experts with four specific technological advances involving a qualitative leap in terms of integrating robotics into everyday life and which are related to the emergence of a new subject, whether they be machines (robots), hybrids (cyborgs), or virtual robots (robots on the net known as intelligent agents) (López Peláez and Kyriakou 2008): the routine incorporation of robotic prostheses in hospitals, daily access to domestic robots that are able to learn the habits of their owners, the use of robots on the net, and the expansion of military or defense robots. All these advances are based on the specific technologies that are forecasted to converge by 2025 in our Delphi studies. The experts' forecast certainty indicates that the development of specific technologies and the robots designed to use them will quickly expand in the coming years and lead to significant social changes in terms of personal interactions, leisure, interaction in the network and security (Table 11.2).

Changes in the technological forecasts, however, do not alter the horizon of 2025 as the moment when automation will take a quantum leap since it is expected that technological problems related to vision, intelligence, language, and mobility will be solved. For both the Japanese case and our research, this time horizon is consistent with Kurzweil's predictions: Biotechnology, robotics, and nanotechnology will reach maturity around 2020, and what he calls the four revolutions will give rise to the singularity, a new post-human reality. Whether or not Kurzweil's singularity hypothesis is true, experts agree that the 2020s will mark a key moment in the reproduction of intelligent human behavior, and the ability of robots to develop behaviors of all types (Kurzweil 2005; Moravec 1999).

Table 11.2 Innovations in robotics and advanced automation by the year 2020

Event	Reference year	Forecast certainty	Social impact of the repercussions
Hospitals will routinely implant robotic prostheses to replace missing limbs in humans	2025	4	4
The percentage of households that will have robots capable of learning or mimetically repeating patterns indicated by their owners will increase	2030	3	3
Robots will be available on the net (intelligent agents)	2020	3	3
Robots will be widely used in the military and by police forces	2020	4	3

Note Forecast certainty: 5 degrees (*1* uncertain, *2* not very certain; *3* certain, *4* quite certain, *5* very certain). Social impact of repercussions: 5 degrees (*1* uncertain; *2* not very certain; *3* certain, *4* quite certain, *5* very certain). *Source* Working Group on Social Trends (GETS 2011)

11.3.2 Always and Everywhere: The Expansion of Industrial and Service Robotics in the Coming Years

In the industrial sector, robots are now a mature, widespread product of increasing versatility. Experts in the automotive sector (Fig. 11.1) predict that 60 % of the activities in the sector will be automated by the year 2015, and that automated tasks will account for 80 % of all activities by 2050. It should be noted that this sector, the main consumer of robots and advanced automation systems, serves as a model of reference. The experts also predict that 70 % of all tasks in other sectors, such as chemicals, petroleum, coal, rubber and plastic products, metal products, and footwear and textiles, will be automated by the year 2050.

By the year 2015, 27.5 % of the activities in the chemicals, petroleum, coal, rubber, and plastics sectors will be automated, while 50 % of activities in the automotive sector will be automated. Two characteristics can be observed in the medium term (2025). First, the increased automation of tasks in all areas of activity; and second, an expansion that will tend to homogenize the levels of automation reached in various areas of industrial activity: 65 % of all jobs in the automotive sector and 42.5 % of all jobs in the footwear and textiles sector will be automated. By 2050, this convergence will be higher, with 60 % of all jobs in the food and beverage sector and 80 % in the automotive sector being automated.

The level of automation predicted by the experts is consistent with the historical expansion of industrial robotics, as well as with the forecasts made by the IFR on the installation of new robots in coming years (IFR 2012a). The increasing automation of tasks in coming years will be the result of three factors: the reduced cost of robots, technological improvements that will permit robots to be used in

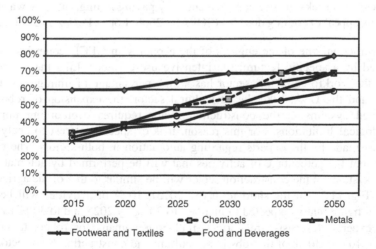

Fig. 11.1 Predictions on the evolution of the percentage of activities that will be automated in various areas of the industrial sector. *Source* López Peláez and Torres Kumbrian (2011)

wider spheres of activity, and the trend to automate an increasing percentage of jobs in the industrial sector.

In the agricultural sector, the experts predict that automated and robotic systems will expand notably in an environment with specific difficulties. While sowing and harvesting require technological advances in terms of mobility, touch, and vision, the automation of classification, packaging and storage systems is relatively easier, but also complex depending on the agricultural product in question. In countries like Italy, Spain, or Greece, agriculture is a sector whose workforce is largely comprised of low-skilled immigrants. This may make it difficult for these workers to find jobs in a sector where the use of agricultural robots is on the rise. The experts predict that 10 % of all agricultural activities will be automated in the short term (by 2015, see Fig. 11.2); a figure that will increase to 25 % by 2025 (when it is predicted that much more powerful robots integrating vision, mobility, language, and intelligence will be developed and used). They also predict that 35 % of all activities in this sector will be automated in the long term (by 2050).

Two trends converge in this process:

- First, the progressive reduction in the price of industrial and service robots (IFR 2012a: 4). The progressively lower cost of robots is associated with other structural advantages of automated and robotic systems: They work 24 h a day, reduce occupational accidents, perform hazardous jobs that put workers' health at risk, and also reduce labor unrest within companies. These advantages enhance the attractiveness of investing in automated robotic systems.
- A second trend that seems to favor the use of a growing number of industrial and service robots has to do with the control of immigration flows. The decision to use robots is supported by the objective difficulties involved in managing the social and economic integration of immigrants. In this sense, the requirements to emigrate to the EU are expected to become tougher, and thus, the economic strategy that seeks greater competiveness by paying immigrants low wages will lose its appeal in coming decades (SOPEMI 2006; López Peláez and Krux 2003).

Construction, one of the engines of the economy in OECD countries, is characterized by a specific feature: the intensive use of labor. Like the agricultural sector, the workforce of this sector has a high percentage of immigrants in both Europe and the USA. Like the agricultural sector, the expansion of robots and automated systems in the construction sector requires overcoming important technological limitations. For this reason, it is of special interest to analyze the forecasts made by the experts regarding automation in both sectors. The experts predict that the percentage of activities that will be performed by automated and robotic systems in the construction sector will be similar to that of the agricultural sector (Fig. 11.2). In the short term (by 2015), 10 % of activities will be automated, a figure that is expected to increase to 20 % by 2025 and to 30 % by 2050. If the experts' forecasts are correct, replacing humans for robots to perform between 10 and 30 % of the jobs in agriculture and construction (two sectors that are characterized by intensive human labor) will have a significant impact on the availability of jobs in these sectors, thus highlighting the need to provide technical training to the workers in these sectors in order to facilitate labor mobility.

Although service robotics is still in an early stage of development, it has remarkable potential. Indeed, in fields such as medical technology, security, and defense, the prototype stage has given way to a wide range of automatic and robotic systems that are routinely used to perform certain activities and tasks (www.ifr.org). Within the service sector, there are four areas that are expected to become highly automated in the coming years, each of which differs in terms of the complexity of automating tasks (Fig. 11.2):

- First, activities related to security, surveillance, and defense. Substantial investments are made in this sector to protect buildings and strategic points, such as railway stations, airports, and nuclear power plants, as well as to reduce the exposure of humans to deadly risks in war-torn environments through the use of automated military vehicles and soldier robots, among other automated systems. The opinions of the experts are clear in this regard: Investments in this type of activities will increase. In the short term (by 2015), technological advances will permit 20 % of the tasks performed to be automated; a figure which will rise to 40 % by 2025, and 60 % by 2050. The idea that security and defense will achieve a level of automation similar to automotive plants indicates the enormous expansion of robots in this sphere of activity; an idea which on the other hand has a strong impact on the collective imagination of the public.
- Second, the automation of tasks in the healthcare sector. Surgical robots have already come to form part and parcel of the daily activities performed by professionals in many hospitals. Spanish experts predict that robotics will expand substantially in this sector, with 10 % of tasks being automated by 2015, 25 % by 2025, and 45 % by 2050. Again, if the forecasts of the experts are correct, in

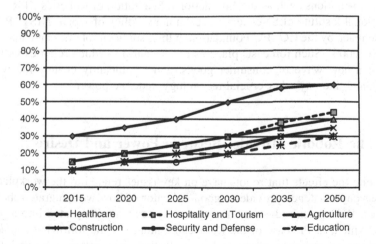

Fig. 11.2 Predictions on the percentage of activities that will be automated in various areas of the agricultural sector and the service sector. *Source* López Peláez and Torres Kumbrian (2011)

the next few years, major changes will occur in the provision of healthcare as the use of robots and advanced automation systems become increasingly widespread.

- The tourist industry accounts for a very high percentage of the GDP of the economies in OECD countries. In this sector, which is characterized by customer relations, the expansion of robotics is an indicator of the technological development of the discipline, as it is possibly the environment with the greatest need for intelligent and mobile robots rather than industrial robots that are designed to perform in a relatively predefined environment. According to the experts, the hospitality and tourism sector will undergo a strong process of automation, thus suggesting that there will be an increase in multifunctional robot prototypes to perform tasks. By 2015, 5 % of the activities performed in the hospitality and tourism sector will be automated. This figure will rise to 15 % by 2025, and to 30 % by 2050, thus reaching higher levels of automation than those found today in sectors such as food and beverages, or the oil industry.
- According to the experts' forecasts, robots will become the colleagues of the future in the fields of education, research, and development. Robots and automated systems will perform 5 % of the activities by 2015, 20 % of the activities by 2025, and 30 % by 2050.

If these predictions are confirmed, the expansion of service robotics will permit automating a number of tasks and activities in the coming years in sectors that traditionally use intensive human labor, thus converting robots into "machines" that are increasingly integrated in the everyday life of twenty-first century citizens. For this reason, we must redefine the classic questions about the impact of robotics and consider new processes of exclusion and inclusion that give rise to a dialectical relationship with robotic innovations. The robotics divide is a probable reality; it is a scenario consistent with the forecasts of experts such as Kurzweil or Levy, and those of institutions such as the International Federation of Robotics (IFR 2012a), the National Institute of Science and Technology Policy of Japan (NISTEP 2005), or those made by the COTEC Foundation in their study on robotics and automation (COTEC 2006). Such forecasts place us in a scenario in which robotics will make way for a new wave of consumer goods, and significantly change the way we behave, and like any other crucial technology, redefine power.

11.4 The Robotics Divide: Between Power and Desire

Examining the effects that robots have on key issues related to the workplace and our everyday existence provides a good indicator of how we integrate robotics in our lives. Robots and advanced automation systems are integrated into a general context which, in the case of the European Union, has the following characteristics: the greater presence of new technologies, the higher intensity of work (which is increasing in all member states as employees must work faster and meet ever tighter deadlines), greater autonomy at work, and an increasing number of highly stressful jobs as demonstrated from the data of five surveys on working conditions in the European Union conducted by the European Foundation for the Improvement of Living and Working Conditions (www.eurofound.europa.eu). As we have seen, working conditions have evolved in many ways. The number of robots and automated systems in the workplace has risen, while organizational models have incorporated technology in a context characterized by increasing job precariousness and greater psychosocial demands (due precisely to the increased intensity of the tasks to be performed): "The subjective indicator of work intensity, which describes workers' experience of high demands, reveals an overall increase in work intensity in most European countries over the past two decades. Although this increase appears to have slowed down since 2005, 62 % of workers in the fifth EWCS report working to tight deadlines (at least a quarter of the time) and 59 % report working at high speed (at least a quarter of the time). Similarly, the proportion of workers whose pace of work is determined by three or more external factors (such as the speed of a machine, client demands, manager) has increased over the past 20 years, though this increase seems to have levelled off since 2005" (Eurofound 2012: 53).

Experts predict that in the next 10 years, the working model in companies with high levels of automation will be quite similar to the model we have today (Table 11.3). In fact, the experts responded in a very similar way to this question in our three Delphi surveys. This suggests that regardless of the training employees receive to work in robotic environments, the job market, the characteristics of the organizations, and labor unrest respond to other issues related to the current socioeconomic model. In this sense, the liberating potential that utopian thinkers have always attributed to robotics does not correspond to the forecasts of the experts. In spite of their potential, robots are integrated into an unstable job market, and as such they will not transform the business model.

Mid-level jobs and career opportunities within companies will remain at levels similar to those of today. In many cases, as one expert warned, mid-level tasks and duties will be automated. The pace of work will be similar to the present one, as will job instability and labor unrest. However, the experts do predict that the number of accidents will be reduced. Thus, the management of automated systems is integrated into a working model and business organization characterized by the increased pace and intensity of work performed by the operators who work with robots. As one expert pointed out, "Workers will always be asked to do more. They will be responsible for performing an increasing number of tasks; even if they work the same hours, the pace will be harder or they will not be able to take breaks." This deterioration in working conditions, however, is not thought to be the result of integrating automated systems, but forms part of the overall deterioration of the job market and working conditions. Indeed, the discourse of the experts clearly points to the fact that today's working conditions are characterized by a lack of continuity in employment, increased transfers and "permanent" job instability. Nonetheless, they believe that employees will be increasingly

Table 11.3 Labor and advanced automation by the year 2020: trends forecasted by the experts

Trend	Delphi study 2002		Delphi study 2005		Delphi study 2011	
	Trend	Forecast certainty	Trend	Forecast certainty	Trend	Forecast certainty
In the next 10 years, will the volume of employment be higher, lower, or the same as now in companies with high levels of automation?	Lower	3	Lower	3	Lower	3
In the next 10 years, will the volume of employment in the service sector be higher, lower, or the same as a result of increased automation?	Same	3	Same	4	Same	3
In the next 10 years, will there be more, less, or the same number of middle management positions in companies with high levels of automation?	Same	4	Less	4	Same	3
In the next 10 years, will there be more, less, or the same job instability in companies with high levels of automation? (job instability refers to poorer working conditions) including the possibility of losing a job	Same	3	Same	4	Same	3
In the next 10 years, will the level of attention needed to operate robots be greater, less or the same as now?	Same	3	Same	4	Less	3
In the next 10 years, will workers operating robots have more, less, or the same amount of control over the work they do? (control refers to the autonomy to decide the pace, sequence and means to perform the task)	Same	4	Same	4	Less	3
In the next 10 years, will the functional mobility of workers who operate robots be greater, less or the same as now? (functional mobility refers to the increased number of tasks assigned to the worker, and changes in the workplace involving higher skills, professional versatility and multi-functionality of the worker)	Greater	3	Greater	4	Same	3

(continued)

Table 11.3 (continued)

Trend	Delphi study 2002		Delphi study 2005		Delphi study 2011	
	Trend	Forecast certainty	Trend	Forecast certainty	Trend	Forecast certainty
In the next 10 years, the saturation level of workers that operate robots will be greater, less or the same as now? (saturation refers to the greater mental effort required to perform a larger number of tasks, greater variation in the tasks, increased responsibility, increasing job uncertainty, longer, more flexible and more irregular working days)	Same	4	Same	3	Same	3
In the next 10 years, will wages (in deflationized euros) be higher, lower, or the same as now in companies with high levels of automation?	Same	3	Same	3	Same	3
In the next 10 years, will the number of occupational accidents in companies with high levels of automation be higher, lower, or the same?	Lower	4	Lower	4	Lower	4
In the next 10 years, will the number of workers who operate robots and suffer from stress be higher, lower, or the same as now?	Same	4	Same	4	Same	4
In the next 10 years, will there be more, less, or the same labor unrest as today in companies with high levels of automation? (labor unrest refers to the number of hours lost due to strikes)	Same	3	Same	4	Same	3

Note Forecast certainty: 5 degrees (*1* uncertain, *2* not very certain, *3* certain, *4* quite certain, *5* very certain). *Source* GETS (2002, 2005, 2011)

specialized, and perform operations in a better and more conscientious manner, while middle management will have fewer but more specialized operators on their staff.

In their forecasts, the experts made two main observations about these developments. First, increased specialization should, theoretically, lead to greater job stability; and secondly, labor costs in companies with high levels of automation should account for an increasingly smaller proportion of the companies' total costs, which in turn would lead to less job instability. However, both job security and wage growth depend on personal variables and the global economic environment. Therefore, while it may seem logical to associate greater special-ization with better job security and higher wages, this is not always the case. In a context in which job stability and wage trends depend on variables related to the global economic environment—whose evolution in recent years has led to the progressive deterioration of working conditions—the greater pace and intensity of work, increased productivity, and the demand for more specialized workers may, paradoxically, have no direct on impact on increased job security and higher wages. Indeed, what happens instead is that instability increases and wage growth does not bear a direct relationship to the increasing demands of the workplace.

In this context, however, automation cannot be understood as the only deter-mining factor, but as one of the many variables in the process to transform work and organizations. The economic logic driving automation, and which automation influences in turn, also responds to other economic, political, and ideological factors which brings us to the limits of global capitalism and its impact on job precari-ousness. The issue of the robotics divide leads to the question about the emerging social model. If, as the experts predict, the current model to manage the new technologies is maintained, we can reasonably expect that robotics, like any other technology, will contribute to reproducing the existing social order and increasing both productivity and inequality between individuals, companies, and states.

Taking into account the basic dimensions of the robotics divide as noted in Chap. 2, it is important to highlight two issues:

- First, in relation to the first two variables (i.e., economic resources and science and technology are needed to develop robotic technology in all areas and reorganize companies, users and civil society in order to increase economic productivity and integrate industrial and service robotics in wider spheres of activity), the experts' forecasts lead us to draw two conclusions: (a) robotics is expanding in all countries and has become a key technology; (b) one of the obstacles to the development of robotics is precisely the inability to reorganize companies and users in order to take full advantage of its potential. As shown in Tables 11.4 and 11.5, the experts have a clear opinion about the factors driving this technology and the obstacles to developing it; both of which will determine whether individuals, companies, and states remain inside or outside the robotics frontier and hence their social and economic exclusion or inclusion.

Table 11.4 Main factors driving the development of robotics in the next 10 years

Driving factors	Absolute value	Weighted value
Increased competitiveness between countries and companies to produce more and better	16	76
Cost savings	11	36
Increasing and improving production	9	26
Introduction of new technologies and robotics in homes	7	24
Support to create businesses in the sector	7	15

Source Working Group on Social Trends (GETS) (2011)

Table 11.5 Main obstacles to the development of robotics in the next 10 years

Obstacles to the development of robotics	Absolute value	Weighted value
The economic crisis	16	78
Lack of investment in $R + D + i$	15	57
Lack of skilled workers	8	18
Technological limitations	8	22
Lack of robots to perform specific tasks	7	21

Source Working Group on Social Trends (GETS) (2011)

- As regards the other two variables (the market economy model and distribution of existing resources in advanced societies, and aspects in which robotics technology leads to a redefinition of power in the military, aerospace, and internet spheres), the experts emphasize that in the coming years, robotics will develop within a socioeconomic system that will remain unchanged. Whether it be with regard to advances in military robotics or the redistribution of resources, it is the current power structures and strategic policies of companies and states that are defining the future model to integrate technology. As has occurred in other periods of history with other crucial technologies, in unequal societies, the effect will be to reinforce the existing power structures. And as a result, people, companies, and countries will increasingly compete to develop more and better robots in strategic sectors such as the military or aerospace.

11.5 Epilogue: From the Digital Divide to the Robotics Divide?

The predictions on the future of our societies arise from a basic principle: people and organizations have the ability to use science and technology in a constructive and advantageous way that also benefits future generations. But this principle also implies exactly the opposite: technologies can be used in an unbeneficial manner. The ability to influence the future, which is the foundation of prospective analyses, goes hand in hand with the role played by science and technology in building the future. Hence, the relevance of experts' forecasts that indicate the technological developments that will materialize in the coming years allows us to make two basic conclusions related to two of the most widely debated issues on the future of robotics: the current model of sociotechnical development and the emergence or not of a human alter ego.

- Firstly, continuing with the sociotechnical trajectory of recent decades, robots and advanced automation systems are being designed and installed in line with the dynamics of neoliberal capitalism in the spheres of industry and leisure, giving rise to a new range of consumer products based on the adaptation of robots to users' needs. One example is the VolksBot or people's robot developed by Fraunhofer and first presented 4 years ago at the Hannover Fair in 2008. VolksBot is a modular system that can be customized to fit the user's needs (www.iuk.fraunhofer.de). Indeed, the expansion of robotics will continue strongly in the industrial sector and increase dramatically in the service sector. In many spheres of non-industrial activity, from agriculture to construction, a large percentage of tasks (over 10 %) are now automated. These technological innovations will be integrated into an economic model characterized by what is known as "permanent instability," a model which easily assimilates the potential of robotics (Lopez Peláez and Kyriakou 2008). Nonetheless, the labor market and business organizations are not expected to be restructured in such a way as to allow for a more equitable redistribution of the wealth generated by a more productive and competitive economy (López Peláez 2007).

- Secondly, and coinciding with the forecasts of Kurzweil and Moravec, crucial advances in robotic technology will be made in the second decade of the century: The convergence of advances in artificial intelligence, language, mobility, and vision will permit more powerful robots to be manufactured. Such advances will significantly increase levels of automation in all sectors and particularly the incorporation of robots in security and defense. From 2025 onwards, robots will be commonplace and the human–machine relationship will intensify in a remarkable manner. Whether in the form of robots, cyborgs or genetically modified human beings that incorporate nanorobots, human life as we know it now will undergo a dramatic process of transformation.

Given that technological innovation will have a growing importance in our lives, we must address the challenges of the coming years. The tasks and activities that can be reorganized and automated in the service sector will tend to be performed by robots (as has happened in the industrial sector). Moreover, the social and political debate on how to manage the impacts of increased interaction between humans and machines in all areas of daily life will take on greater importance. As the various technological trajectories throughout history have shown, there is no single possible destination, not all social groups "win," and a key technology often implies that one group, nation or civilization dominates over another (as the history of military technology attests to).

When analyzing the main obstacles to the development of robotics and advanced automation, the experts we interviewed for our study coincided with other studies in noting a key problem in advanced democratic societies: the social changes that must be undertaken to achieve a correct balance between technological development and rising unemployment (López Peláez and Torres Kumbrián 2011). This is a controversial issue, and authors such as Castells have repeatedly shown that there does not exist a structural relationship between the incorporation of technologies and rising unemployment (Castells 1996). In our opinion, the key question is that the analysis of trends in the evolution of technology allows us to objectify the current model of development in debates on a possible future. By doing so, we have a greater ability to act upon our own sociotechnical environment. While this is one of the greatest contributions of prospective research on key technologies, it is still in an early stage of development. In this regard, one of the main measures public administrations should take to promote the development of robotics in the coming years is, as the experts state, "to ensure that such development benefits society" (López Peláez 2007: 394).

Moving beyond the discussions on the digital divide, it is possible that a new social and technological divide will occur, the robotics divide; a divide which could have drastic consequences on our way of life, as we have shown throughout this book. It is also possible that we might move toward a more integrated and cohesive society due to the increased productivity and efficiency of automatic and robotic systems. Given this open horizon in which the probable future depends on the decisions we make in the present, it is essential that prospective studies become a basic tool in our advanced societies. Nonetheless, what is at stake is the dominant political model that is already shaping our sociotechnical model of development; an issue that comes to light when analyzing the forecasts of experts. In this sense, when we open the black box of our technological trajectories, we are obligated to analyze the future that we are in the process of creating. Although we try hard to forget, to bury it in a naïve technological optimism, our present always reveals the processes of inequality in which we are immersed. In the face of such inequalities, we must take awareness of our probable future, assess its consequences, and make the right decisions to ensure that another future is possible beyond the neo-Darwinian approaches characteristic of the neoliberal sociotechnical model. This discussion aims to make a modest contribution to this collective work on the robotics divide.

References

Adler M, Ziglio E (1996) Gazing into the oracle. The Delphi Method and its application to social policy and public health. Jessica Kingsley Publishers, London

Aibar E (2011) Usuarios y tecnologías de la información: de la administración electrónica al software libre. In: González de la Fe T, López Peláez A (eds) Innovación, conocimiento científico y cambio social. Ensayos de sociología ibérica de la ciencia y la tecnología. CIS, Madrid, pp 163–190

Bijker W (1995) Of bicycles, bakelites, and bulbs. Toward a theory of Sociotechnical change. MIT Press, Cambridge

Castells M (1996) La era de la información. Economía, sociedad y cultura, vol 1. La sociedad red. Alianza Editorial, Madrid

Collingridge D (1980) The social control of technology. Pinter, London

COTEC (2003) Tendencias tecnológicas en Europa. Análisis de los procesos de prospective. Fundación COTEC para la innovación tecnológica, Madrid

COTEC (2006) Robótica y automatización 2006. Fundación COTEC para la innovación tecnológica, Madrid

Eurofound (2012) Fifth European working conditions survey. Publications Office of the European Union, Luxembourg

FraunhoferInstitut Systemtechnik und Inovationsforschung (FISI) (1998) Delphi 98. Studie zur Globalen Entwicklung von Wissenschaft und Technik. Methoden und Dateband. Bundesministerium für Bildung, Wissenschaft, Forschung und Technologie, Detuschland

GETS (2002) Delphi study on scientific and technological trends. Sistema, Madrid

GETS (2005) Delphi study on scientific and technological trends. Sistema, Madrid

GETS (2011) Delphi study on scientific and technological trends. Sistema, Madrid

Goldschmidt P (1975) Scientific inquiry or political critique? In: Sackman H (ed) Remarks on Delphi assessment, expert opinion, forecasting, and group process. Technological Forecasting and Social Change, vol 7. pp 195–213

Häder M, Häder S (eds) (2000) Die DelphiTechnik in den Sozialwissenschaften. Methodische Forschungen und innovative Anwendungen. Westdeutscher Verlag, Wiesbaden

Institut für Trendanalysen und Krisenforschung (ITK) (1998) Delphi Report Austria, vol 5: Gesellschafts und KulturDelphi I: Die Zukunft der österreichischen Gesellschaft. ExpertenSzenarien für die Jahre 2003-2013-2028. Bundesministerium für Wissenschaft und Verkehr, Wien

International Federation of Robotics (IFR) (2012a) World Robotics. Industrial Robots 2012. IFR, Geneva

International Federation of Robotics (IFR) (2012a) World Robotics. Service Robots 2012. IFR, Geneva

Keenan M, Butter M, Sainz de la Fuente G, Popper R (2006) Mapping foresight in Europe and other regions of the world. Highlights from the annual mapping of the EFMN in 2005–2006. European Commission DG Research, Brussels

Kurzweil R (2005) The singularity is near. When humans transcend biology. Viking Penguin Group, New York

Linstone HA, Turoff M (eds) (1975) The Delphi method. Techniques and applications. AddisonWesley, Reading (Mass)

López Peláez A (2000a) Impactos de la Robótica y la Automatización Avanzada en el Trabajo: Estudio Delphi. Sistema, Madrid

López Peláez A (2000b) Towards a new work pattern? Trends of automation and robotics systems in manufacturing and services. Robot Newslett (J Int Fed Robot) 40:810

López Peláez A (2007) Innovación tecnológica, crecimiento económico y automatización avanzada: paradojas de la globalización. In: Tezanos JF (ed) Los impactos de la revolución científico tecnológica. Noveno Foro sobre Tendencias Sociales. Sistema, Madrid, pp 355–400

López Peláez A, Díaz Martínez JA (2007) Science, technology and democracy: perspectives about the complex relation between the scientific community, the scientific journalist and public opinion. Soc Epistemol 21(1):55–68

López Peláez A, Krux M (2000) Social impacts of robotics and advanced automation towards the Year 2010. The IPTS Rep (edited by The Institute for Prospective Technological Studies—European Commission) 48:34–40

López Peláez A, Krux M (2002) Future trends in health and safety at work: new technologies, automation and stress. The IPTS Rep (edited by The Institute for Prospective Technological Studies—European Commission) 65:24–33

López Peláez A, Krux M (2003) New Technologies and new migrations: strategies to enhance social cohesion in tomorrow's Europe. The IPTS Rep (edited by The Institute for Prospective Technological Studies—European Commission) 80:11–17

López Peláez A, Kyriakou D (2008) Robots, genes and bytes: technology development and social changes towards the year 2020. Technol Forecast Soc Chang 75:1176–1201

López Peláez A, Segado Sánchez-Cabezudo S (2009) Transporte, trabajo y salud: perspectivas sociológicas sobre la liberalización del transporte ferroviario. Sociología del Trabajo 67:151–173

López Peláez A, Segado Sánchez-Cabezudo S (2010) Privatization policies or degradation policies? The case of Spanish railways. Revista Internacional de Sociología (RIS) 68(3):757–773

López Peláez A, Torres Kumbrian R (2011) Cyborgs, automatización avanzada y cambio socia In: González de la Fe T, López Peláez A (eds) Innovación, conocimiento científico y camb social. CIS, Madrid, pp 191–214

López Peláez A, Segado Sánchez-Cabezudo S, Kyriakou D (2012) Railway transpc liberalization in the European Union: freight, labour and health towards the year 2020 Spain. Technol Forecast Soc Chang 79:1388–1398

Loveridge D, Georghiou L, Neveda M (1995) United Kingdom technology foresight programm Delphi survey. The University of Manchester, Manchester

Ministére de l'Enseignemetn Supérieur et de la Recherche (MESR) (1995) Enquête sur l technologies du futur par la méthode Delphi. MESR, Paris

Moravec H (1999) Robot. Mere machine to transcendent mind. Oxford University Press, Ne York

National Institute of Science and Technology Policy (NISTEP) (2002) The seventh technology foresight. Future technology in Japan toward the Year 2030. Ministry of Education, Culture, Sports, Science and Technology, Tokyo

National Institute of Science and Technology Policy (NISTEP) (2005) Science and technology foresight survey. Delphi analysis. NISTEP report no 97. Ministry of Education, Culture, Sports, Science and Technology, Tokyo

Orlikowski W (2000) Using technology and constituting structures: a practical lens for studying technology in organizations. Organ Sci 11(4):404–428

SOPEMI (2006) International migration outlook. OECD, Paris

Working Group on Social Trends (GETS) (1996) Delphi study on scientific and technological trends. Sistema, Madrid

Working Group on Social Trends (GETS) (2011) Delphi study on scientific and technological trends. Sistema, Madrid

Book Description

Technology and power are closely intertwined. As the specialized literature demonstrates, technologies can reinforce processes of exclusion in contemporary societies, or permit new strategies for social inclusion to be developed. This book examines a newly emerging technological divide: the robotics divide. From different theoretical perspectives, the authors of the book address key issues related to the expansion of robotics technology, its gradual implementation in the service sector, leisure and domestic life, as well as new applications in the social services field and for the disabled. In exploring the changing trends in robotics in the coming years, the authors focus on the sociotechnical model of development in which we are immersed, the likelihood of a new robotics divide in sectors such as space or the military, and the day-to-day relationship between human beings and machines. Through an innovative and interdisciplinary theoretical approach which proposes new concepts like the robotics divide, the different chapters of the book permit us to open the black box of technological development in a key area in which humans are creating an alter ego (the robot); a phenomenon which will redefine the distribution of power in societies of the twenty-first century.

A. López Peláez (ed.), *The Robotics Divide*,
DOI: 10.1007/978-1-4471-5358-0, © Springer-Verlag London 2014

CV Antonio López Peláez

Antonio López Peláez holds a Ph.D. in Philosophy and Sociology. He is a full professor of Social Work and Social Services at the Department of Social Work of the Faculty of Law at the UNED, the largest public university in Spain. His research interests include the analysis of social trends, methods of social intervention, and the intersections of new technologies and social work. He has been a visiting scholar at the School of Social Welfare of the University of California at Berkeley (USA), Universität Potsdam (Germany), the Department of Social Work at Universidad de Sonora (Hermosillo, Mexico) and the Universidad Americana (Managua, Nicaragua).

In 2001, he received the award for excellence in doctoral research for his Ph.D. in Sociology. In 2008, his article (with D. Kyriakou) "Robots, Genes and Bytes: Technology Development and Social Changes Towards the Year 2020" was named one of the Top 25 Hottest Articles in Technological Forecasting and Social Change. His most recent publications in the field of futures studies include López Peláez, Segado Sánchez-Cabezudo S, Kyriakou D (2012) Railway transport liberalization in the European Union: Freight, labour and health towards the year 2020 in Spain. Technological Forecasting and Social Change 79: 1388–1398; López Peláez A., Segado Sánchez-Cabezudo S (2010) Liberalization Policies or Degradation Policies? The Spanish Railway Case. Revista Internacional de Sociología (RIS) 68: 757–773; and López Peláez A (2009) Prospective and Social Change: How to Direct Research and innovation Policies in Technologically Advanced Societies. Arbor. Ciencia, Pensamiento y Cultura 738: 825–836.

A. López Peláez (ed.), *The Robotics Divide*,
DOI: 10.1007/978-1-4471-5358-0, © Springer-Verlag London 2014

Printed in the United States
By Bookmasters